IMAGES
of America

ST. PAUL'S PARISH

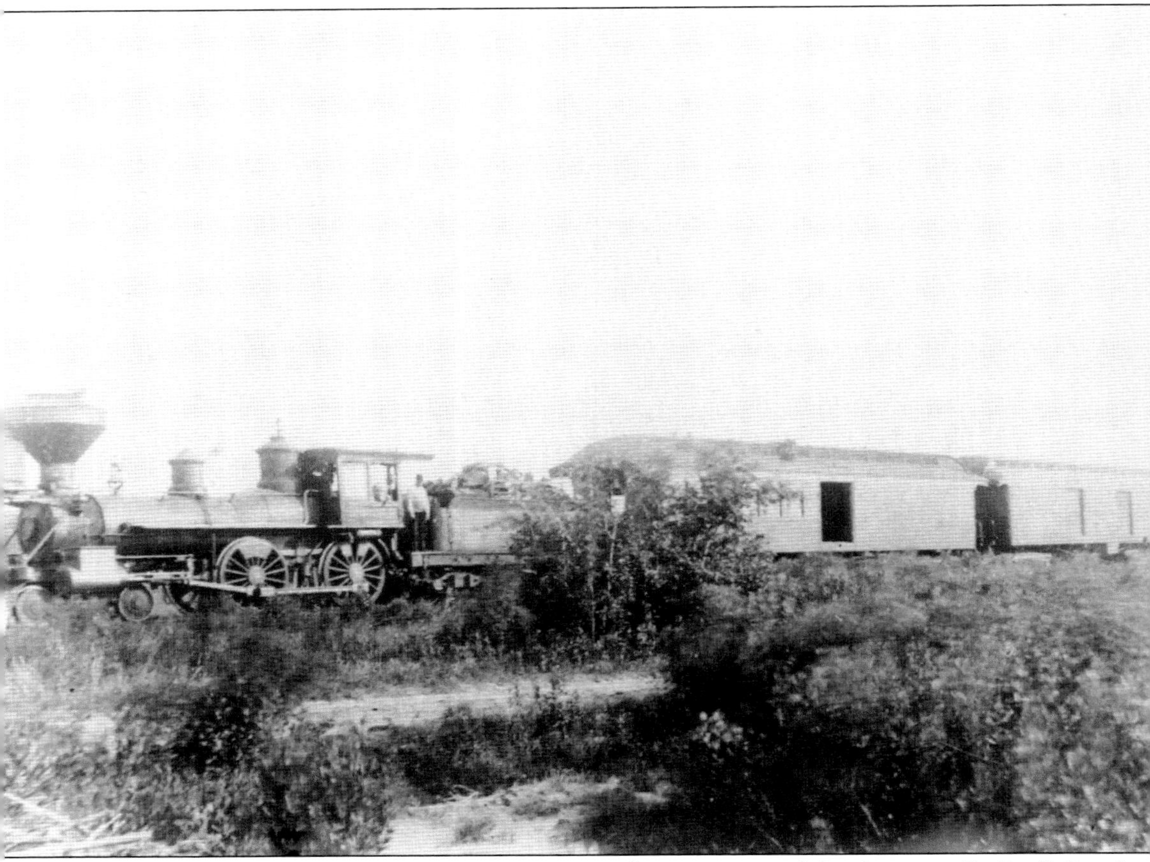

Seen here is a picture of a steam locomotive pulling into one of the depots in St. Paul's Parish around the turn of the 20th century. Railroads would be key to the development of the area known as the "Cabbage Capital of the World." (Courtesy of Buddy Hill and Rita Sanders.)

On the Cover: A young African American boy stands alongside a train that is pulling into what is believed to be Meggett Depot. He is surrounded by crates that are full of produce, namely cabbages. (Courtesy of the Town of Meggett, South Carolina.)

IMAGES
of America

ST. PAUL'S PARISH

Jennifer H. Gilliland
Foreword by John "Jackie" Boineau

ARCADIA
PUBLISHING

ISBN 978-0-7385-9119-3

Published by Arcadia Publishing
Charleston, South Carolina

Printed in the United States of America

Library of Congress Control Number: 2011935452

For all general information, please contact Arcadia Publishing:
Telephone 843-853-2070
Fax 843-853-0044
E-mail sales@arcadiapublishing.com
For customer service and orders:
Toll-Free 1-888-313-2665

Visit us on the Internet at www.arcadiapublishing.com

This book is dedicated first and foremost to my husband, Eric, and my daughter, Piper Leigh. This project would not have been possible without their support, encouragement, and understanding. I would also like to thank all of the people here in the parish who gave of their time and their photographs.

Contents

FOREWORD

St. Paul's Parish begins about 10 miles outside of the city limits of Charleston, South Carolina, and stretches westward out toward Edisto Island. The parish was officially established in 1706 when the Lords Proprietors divided the Carolinas into counties and then divided them into squares and baronies. St. Paul's Parish was originally part of Colleton County and remained there until the early 1900s. Rich in history, the parish is comprised of a proud people with deep ties to the land. Early settlers began to farm the land in the mid-17th century and even founded the second-oldest settlement in South Carolina, Willtown (or New London), in 1685.

Throughout the years, the small towns that make up St. Paul's Parish—Hollywood, Adams Run, Meggett, Yonges Island, Ravenel, and Rantowles—have evolved and changed through wars, depressions, recessions, segregation, desegregation, and booms and busts. The parish has been the scene of historic battles dating to the Revolutionary War, the birthplace of governors, the site of a Confederate hospital, and even the "Cabbage Capital of the World." Perhaps most importantly, the families that helped settle the area as well as the families that reside their today simply think of it as home—a place where their great-grandparents, grandparents, mothers, fathers, aunts, uncles, cousins, and friends have always lived and continue to live today.

What has stayed the same throughout the years is something intangible that residents can only describe as a place "close to heaven"—where time has stopped, where neighbors know each other and help each other out, and where there is a love of the land and a desire to preserve the heritage in this little-known part of the Lowcountry.

—John "Jackie" Boineau

ACKNOWLEDGMENTS

This book would not have been possible without the help of the residents of St. Paul's Parish who welcomed me into their homes, took time to look through family memorabilia, and shared stories from the past. Thanks go to the Town of Meggett, Anna Blythe, Caroline Jarvis, Linda Allston, Patrick Butler, Ronald Rash, Lisa and David Kennedy, Marsha Blandenburg, Susan Stevens, Joyce "Dumpy" Towles Barnwell Cave, John Waring, Bobby Shealy, Mona Butler Williams, Judy and Reeves Bair, Buddy Hill, Jackie Boineau, Anne Sanders Groover, Rosanne King Jones, Constance King Puckett, Elizabeth Waters, Wilbur Jones, Bobby Garvin, Natalie Towles Whiteman, Liz Whiteman, Beth Van Norte, Anne Commins, Sarah Roberts, William "Billy" Geraty, and Metal Trades Inc.

INTRODUCTION

St. Paul's Parish was established by the Church Act of 1706 that divided the colony of South Carolina into parishes. Colleton County, which contained all of St. Paul's Parish, was officially divided into St. Paul's and St. Bartholomew's. St. Paul's was bounded to the northeast by the Stono River and what was then Berkeley County, to the southeast by the Atlantic Ocean, and to the west by the South Edisto River. The parish was divided again in 1734, and many of the larger barrier islands (Johns Island, Wadmalaw Island, Edisto Island) and some other smaller islands were organized as a different parish—St. John's. Island, Wadmalaw Island, Edisto Island) and some other smaller islands were organized as a different parish—St. John's. Today, the parish remains relatively unchanged from the 1734 division. Oftentimes, the people, events, and places of this area, which is now part of Charleston County, are overlooked, given the natural attention that many historians and visitors focus on the nearby famous historic city of Charleston. . Historically speaking, the towns that comprise the parish—Hollywood, Rantowles, Adams Run, Meggett, Yonges Island, and Ravenel—have some significant history in and of themselves. Adams Run is a small pineland village about five miles from what was Willtown, the second-oldest settlement in South Carolina. It was a prominent summer village in the Lowcountry and was settled mainly by the planter gentry and their families. Rantowles was named for Alexander Rantowle, an early landowner. Hollywood was a town that sprang up during the railroad development in the parish in the early 20th century. Legend has it that it was named after Hollywood Plantation. Both Meggett and Yonges Island were named after local plantation owners, William Meggett and Francis Yonge.

St. Paul's Parish can claim fame for many things, including the second-oldest settlement in South Carolina being founded here, and the Stono Rebellion slave uprising occurring in 1739. It is the birthplace of state governors, the site of Revolutionary and Civil War battles and encampments, and it has many historic plantation homes. It is a place of inspiration for artists as well as authors and is considered the "Cabbage Capital of the World." The parish is an example of how agriculture and industry flourished in the late 19th and early 20th centuries with the expansion of the railroads. The railroads allowed people to use the land as well as the water to more easily move goods, thus crating wealth and enabling upward mobility both economically and socially. Overall, the story of the parish and its people showcases the rise and fall of small towns over the span of over 150 years.

The first chapter, "Agriculture and Industry," features people and places that helped put the towns of the parish on the map. From the early rice and cotton plantations to the truck farms of the early 20th century and from the industrial processing of oysters to the locally owned shops, car dealerships, cafés, and general stores, there are photographs that will jog the memory of a time and place that is no longer. It is a chapter that tells the story of how small, rural towns grew into significant centers of economic activity.

Entitled "Gathering Places," the second chapter showcases all of the places where the people of the parish joined together for fun, work, worship, and education. The photographs capture the essence of what it was like to be a child, parent, teacher, preacher, grandparent, or neighbor over the course of nearly two centuries in a small, Southern town. These pictures represent people and places that are no longer with us as well as others that have moved on to a different season of life; all have a story to tell, a memory to share, a desire to capture the past, or a bit of captured past to communicate.

The third chapter, "Trains, Planes, and Automobiles," is a series of pictures that helps recall what it was like to ride the train or how it felt to wait for an out-of-town visitor to arrive by rail. More importantly, the railroads were instrumental to the development of the parish, and the numerous images of trains, depots, and tracks showcase what an integral part the railroads played in the lives of parish residents. Without the railroads, the towns of the parish would not have flourished the way that they did. Automobiles came of age in the early 20th century and were prized possessions of St. Paul's Parish residents just as they were for most Americans at that time. Air travel was not prevalent back then, but just imagine for a moment how frightening it would have been to see a plane crash in your town in the 1920s or how rare it might have been to capture the crash on film.

Finally and perhaps most importantly, the fourth chapter, "Parish People," is a series of pictures dedicated to the people of St. Paul's Parish. These are the cherished images of our grandparents, parents, teachers, friends, and neighbors, many of whom remain in the parish today. The pictures will make you laugh, cry, and smile as you relive the memories of yesteryear. There are numerous people over the years that have made St. Paul's Parish what it is today, and this area has been blessed with a large number of distinguished inhabitants. While this chapter strives to represent as many people and families in the parish as possible, space, availability of information, and photographic quality have determined the contents of this section.

St. Paul's Parish and the individual towns that make up the parish remain vibrant communities today, with many new families relocating and choosing to live in this "suburb" of Charleston. Families appreciate the open spaces, the sense of community, and the friendliness of neighbors, which can be so hard to find in today's society. The residents of the parish value their place in both American and South Carolina history and are proud to share their pictures and stories in the hopes that everyone—whether a visitor to the area, a new resident, or "lifer"—can develop an appreciation of this part of the Lowcountry.

One

Agriculture and Industry

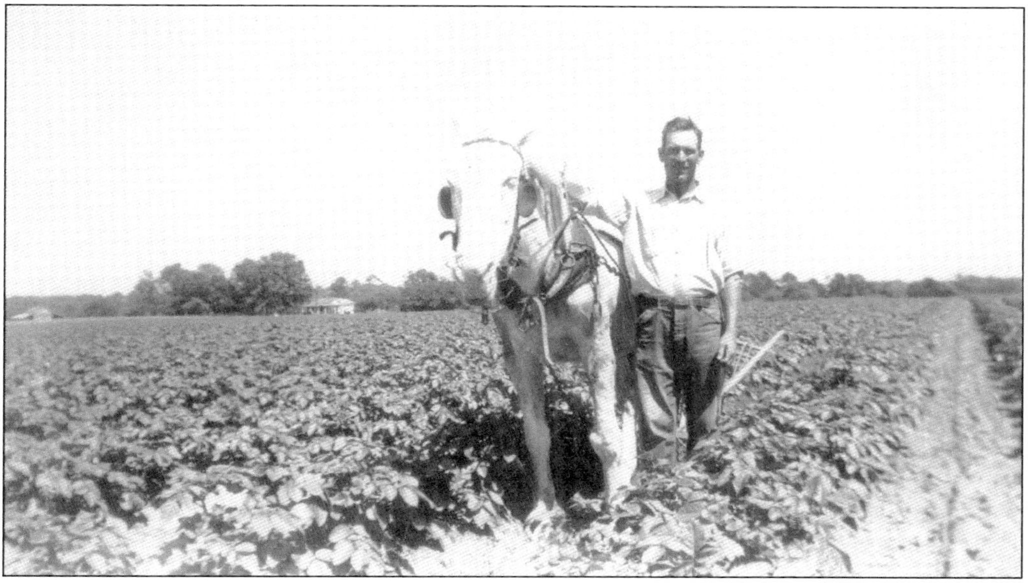

Agriculture and industry played a large role in the development of St. Paul's Parish as an economically viable area outside of Charleston, South Carolina. From rice and cotton to the evolution of truck farming or the processing of oysters at the local oyster factory, the parish encouraged, developed, and supported various industries over the years. Like many other farmers in the parish, E.B. Prentiss relied heavily on his farm animals to help raise successful crops. Prentiss is seen around 1946 with his mule Alice in the field at his farm, Cherry Tree, off Toogoodoo Road. (Courtesy of Caroline Jarvis.)

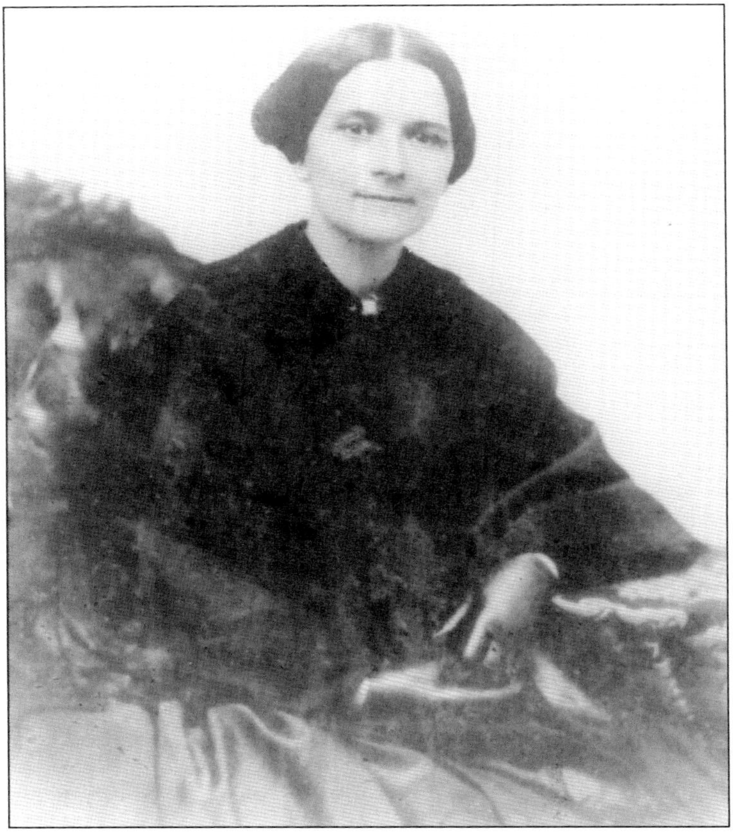

Regina Margaret Coachman Allston, daughter of Francis Green Coachman, and her husband, John Elias Allston (1820–1856), bought Hermitage Plantation from William Wescoat in the 1850s. In 1880, the Hermitage tract had 90 acres cultivated in rice. In 1875, Regina, who was now widowed, purchased Fairview Plantation from John Marshall. When the property was auctioned off at a sheriff's sale, portions of that land were cultivated for rice as well. Fairview Plantation is located just north of Slann's Island in the fork of Edings Creek and Toogoodoo Creek and is still owned by Allston's descendants. (Both, courtesy of Linda Allston.)

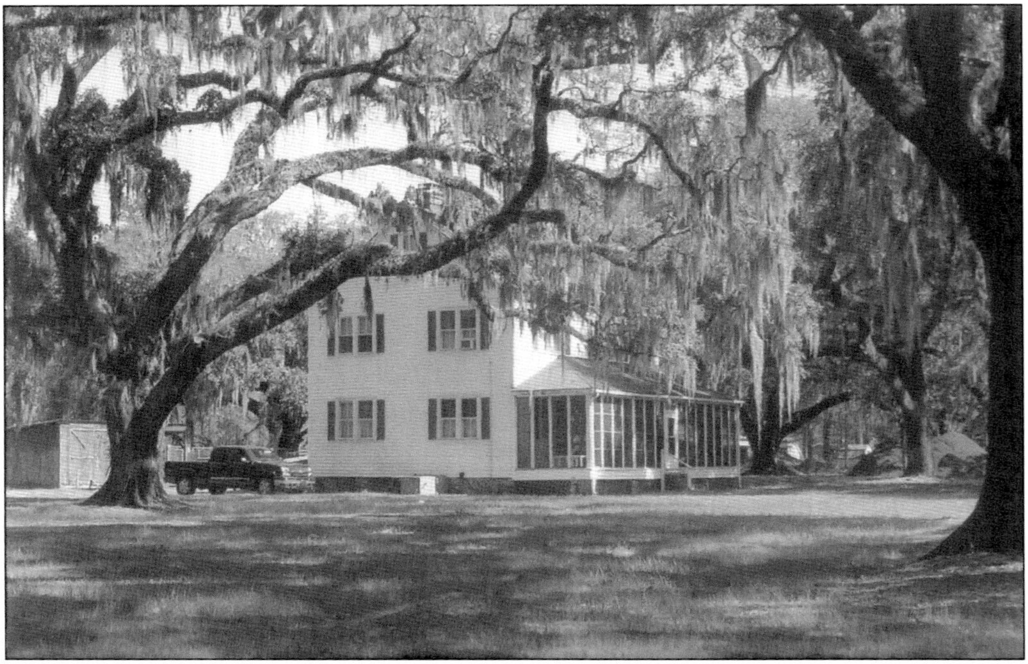

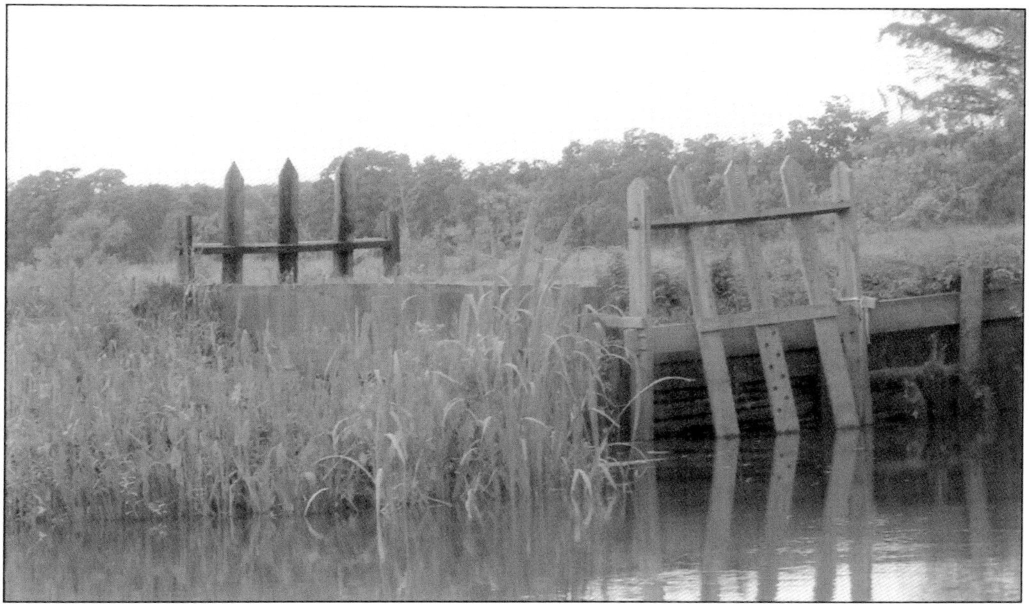

Colonists successfully grew small amounts of rice in the Lowcountry by 1690. During the period 1820–1860, between 40 million and 100 million pounds of rice were exported annually. Rice made South Carolina one of the wealthiest states in America. Rice "trunks," or flood-control gates, were built by skilled artisans and engineers to control the flow of water to the rice fields. Skillful operation of these locks and trunks allowed freshwater to cover the fields at proper times and prevented saltwater from damaging the crops. Seen in the photograph above is one such rice trunk that still exists on the South Edisto River today. Below are the remains of a rice plantation along the river. (Both, courtesy of Linda Allston.)

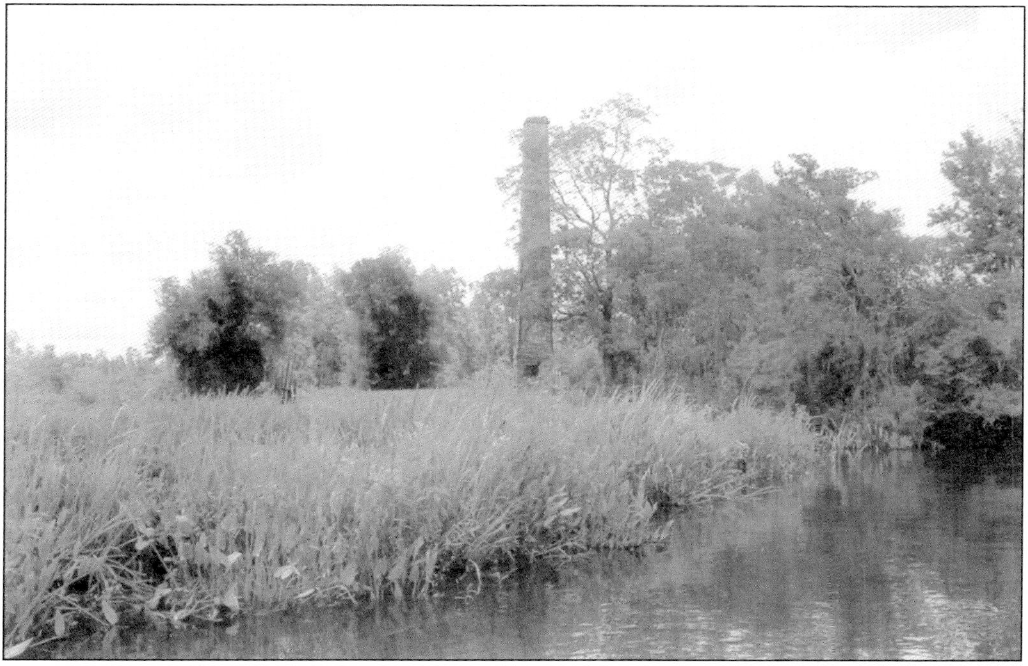

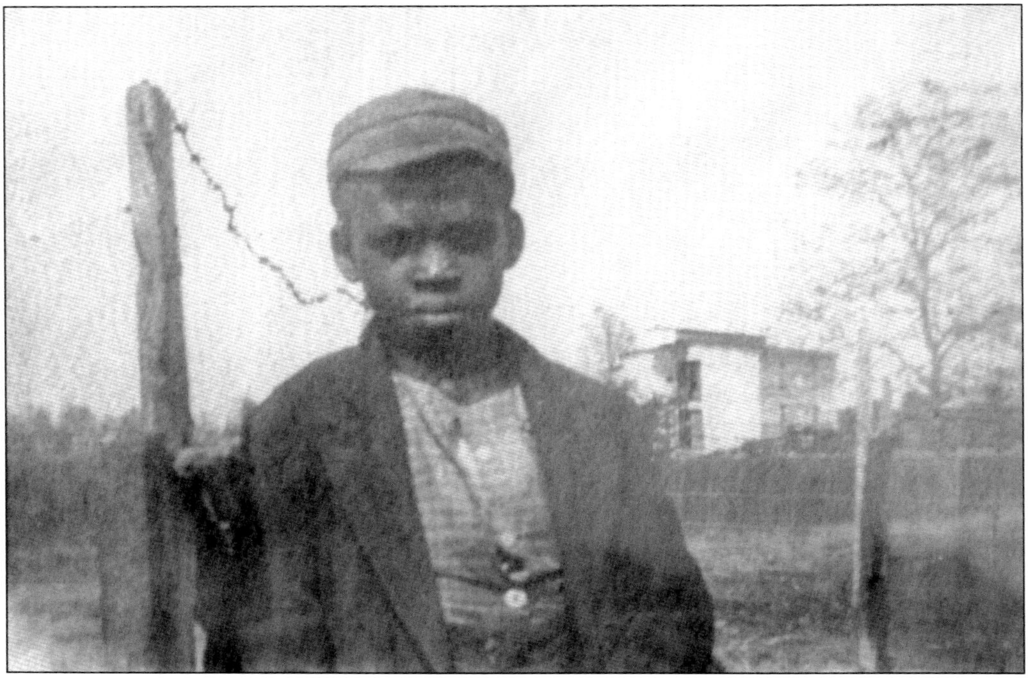

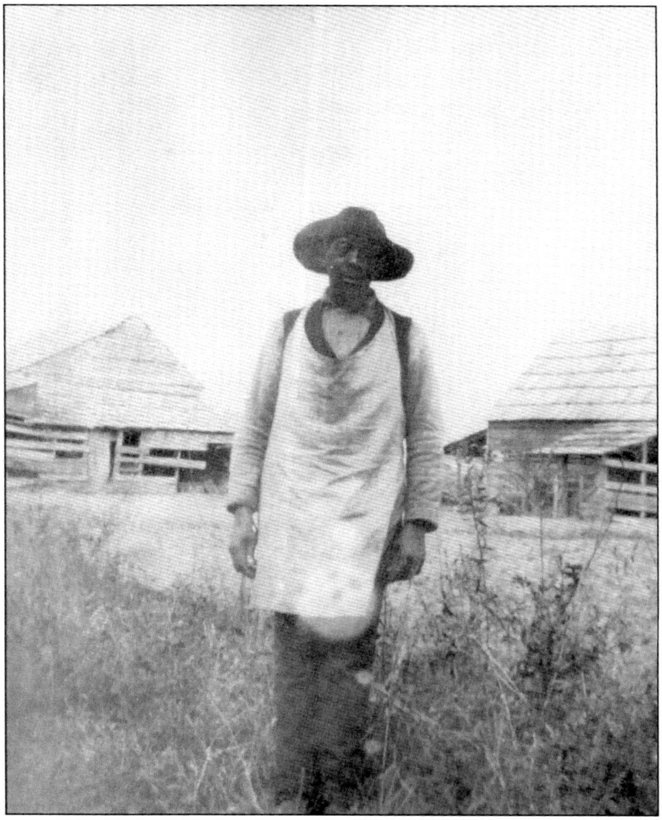

In the picture above, taken on February 17, 1909, is a young African American boy (approximately six years in age) named William, who lived and worked on Rescue Plantation, which was a large farming operation owned by the Behling family in Meggett, South Carolina. In the December 1908 picture at left is an African American man called "Wee Willy," who also worked on Rescue Plantation assisting the Behling family with the planting and harvesting of cash crops, such as cabbages, radishes, and potatoes. (Both, courtesy of Sarah Roberts.)

In this photograph, taken on April 23, 1947, are Jennie "May" Tumblestone Prentiss and her son Burton Prentiss in front of the packing shed on their farm off Toogoodoo Road. The Prentiss family raised cabbages, beans, bell peppers, and potatoes to take to market and also had a fruit orchard on the property, hence the name Cherry Tree Farm. (Courtesy of Caroline Johnson.)

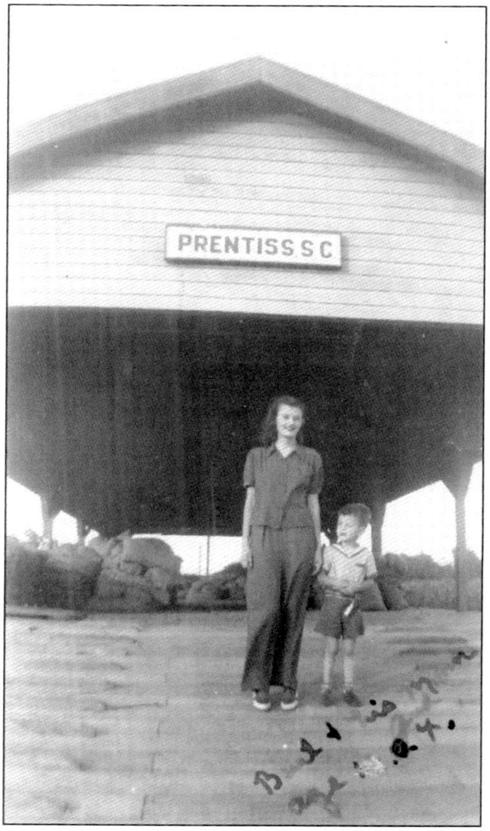

Seen standing in front of the Behling family packing shed are, from left to right, Abigail Sarah "Ms. Abbie" LaRoche, her daughter Liz Behling, and Luder Cadow Behling. The Behlings owned and operated Rescue Plantation in Meggett and ran a family lumber business. (Courtesy of Pat Butler and Sarah Roberts.)

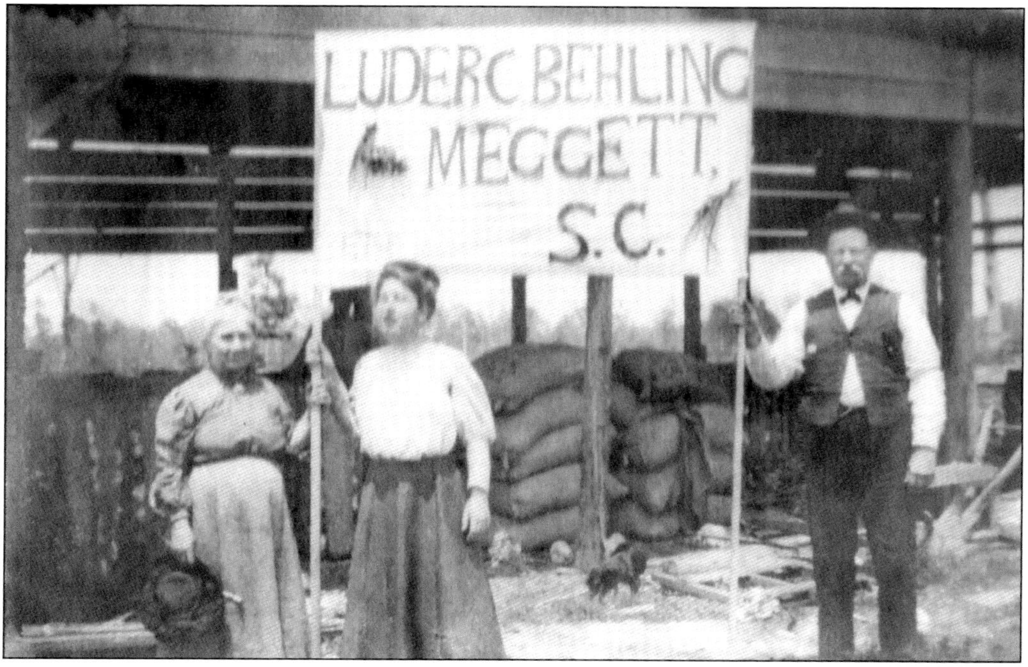

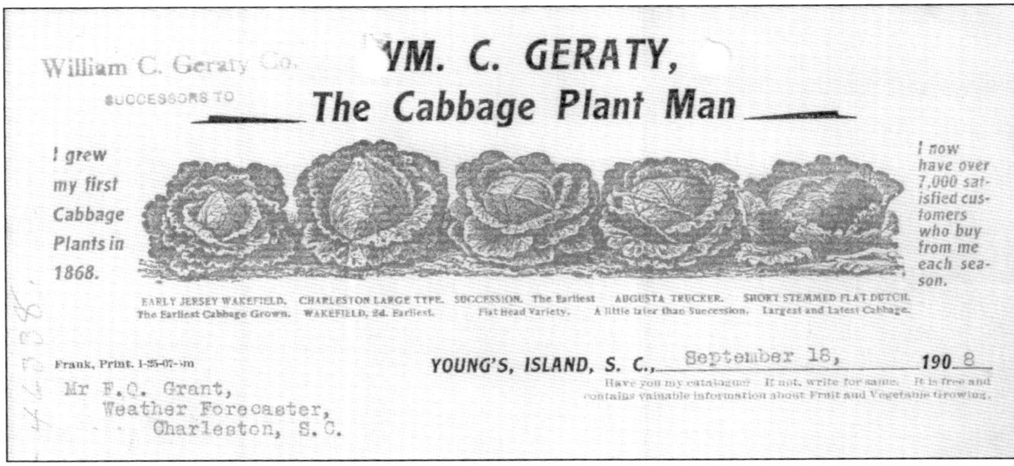

William C. Geraty Co.
SUCCESSORS TO
WM. C. GERATY,
The Cabbage Plant Man

I grew my first Cabbage Plants in 1868.

I now have over 7,000 satisfied customers who buy from me each season.

EARLY JERSEY WAKEFIELD. The Earliest Cabbage Grown. CHARLESTON LARGE TYPE. WAKEFIELD, 2d. Earliest. SUCCESSION. The Earliest Flat Head Variety. AUGUSTA TRUCKER. A little later than Succession. SHORT STEMMED FLAT DUTCH. Largest and Latest Cabbage.

YOUNG'S, ISLAND, S. C., September 18, 1908

Have you my catalogue? If not, write for same. It is free and contains valuable information about Fruit and Vegetable Growing.

Mr F.Q. Grant,
Weather Forecaster,
Charleston, S.C.

The 1908 letterhead of William Christopher Geraty advertises his company and a variety of cabbages: Early Jersey Wakefield, Charleston Large Type, Succession, Augusta Trucker, and Short Stemmed Flat Dutch. (Courtesy of William "Billy" Geraty.)

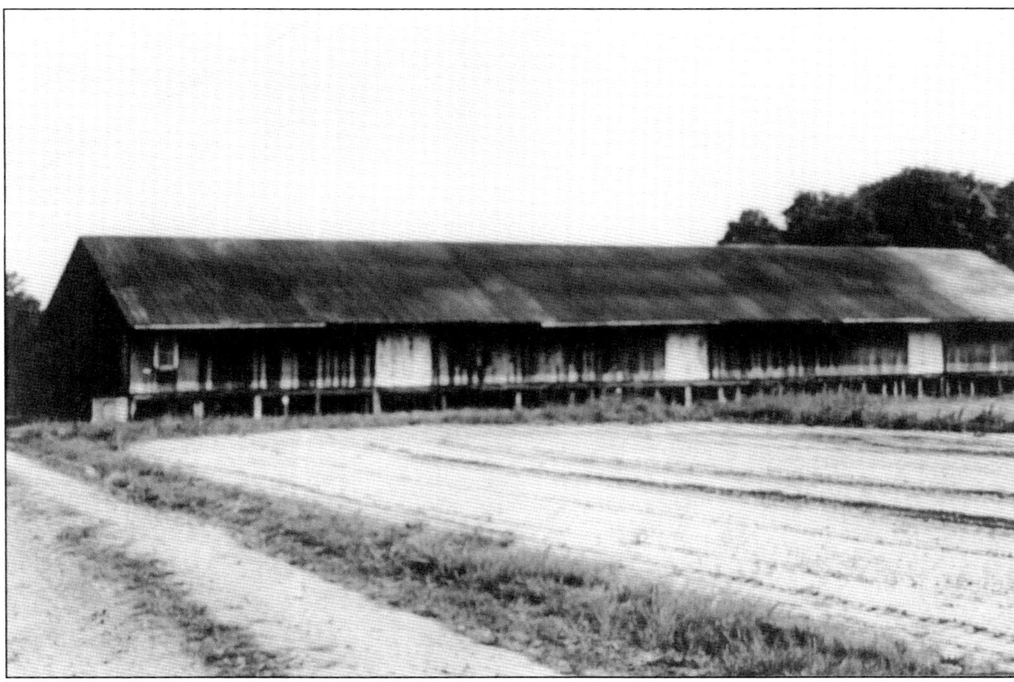

This early-1970s photograph shows the packing shed belonging to the Gibson family. It is located off what is now the intersection of Ethel Post Office Road and Quigley Road. Produce sheds were built along the main line and spurs of the railroads to store and load farmers' produce in an effective manner. The majority of these sheds, which once dotted the landscape, are no longer standing or are in poor condition. (Courtesy of John Waring.)

In the photograph at right are Weston Harrison Geraty (left), otherwise known as "Duck," and John Gibson, brother of Charles Gibson. The Gibsons were a prominent farming family in the Yonges Island area. According to a *Charleston News and Courier* article from 1912, C.M. Gibson farmed 40 acres on Yonges Island, which produced 12 million cabbage plants, and S.M. Gibson farmed 16 acres in Meggett, which produced 7 million cabbage plants. Below is the letterhead for S.M. Gibson dated July 5, 1922, showcasing frost-proof produce, including four different varieties of cabbages. (Right, courtesy of William "Billy" Geraty; below, courtesy of Mona Butler Williams.)

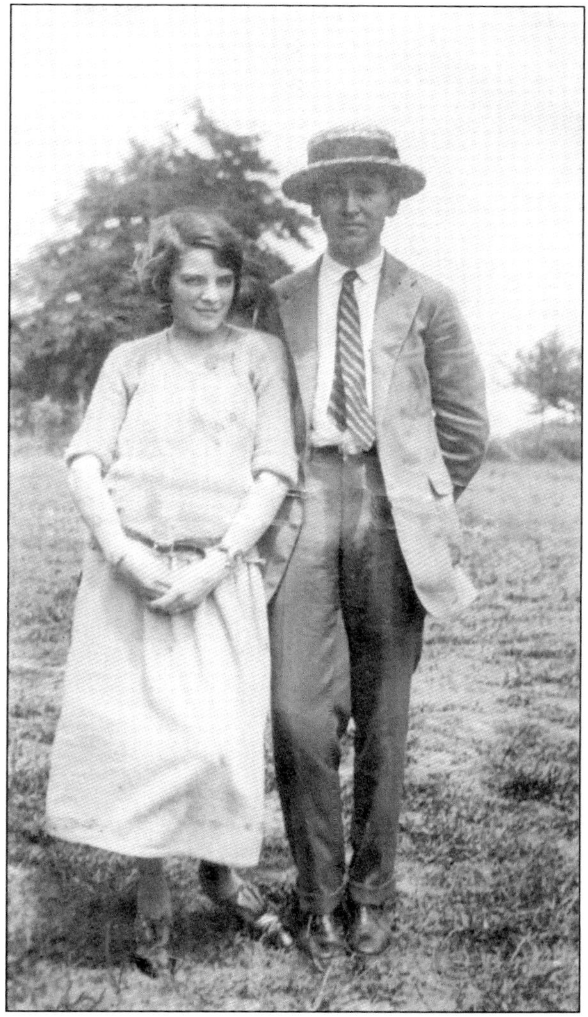

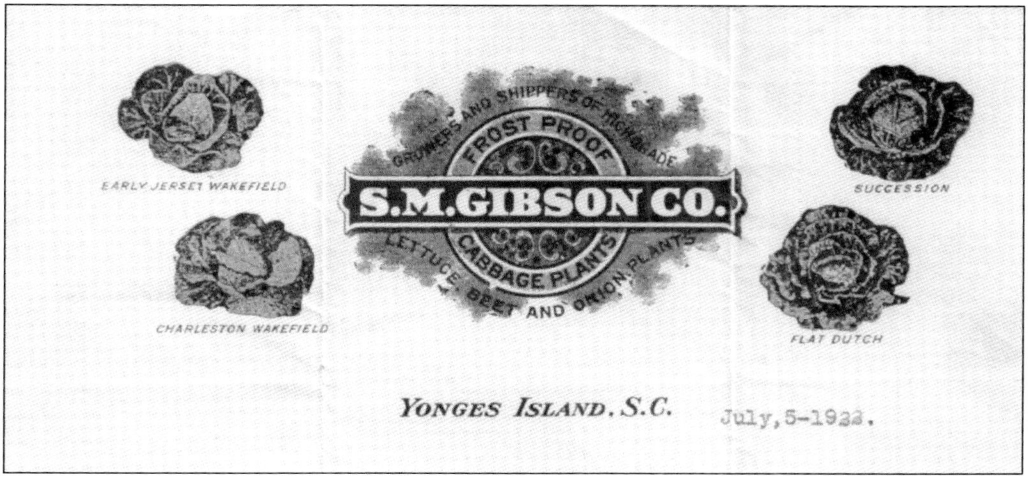

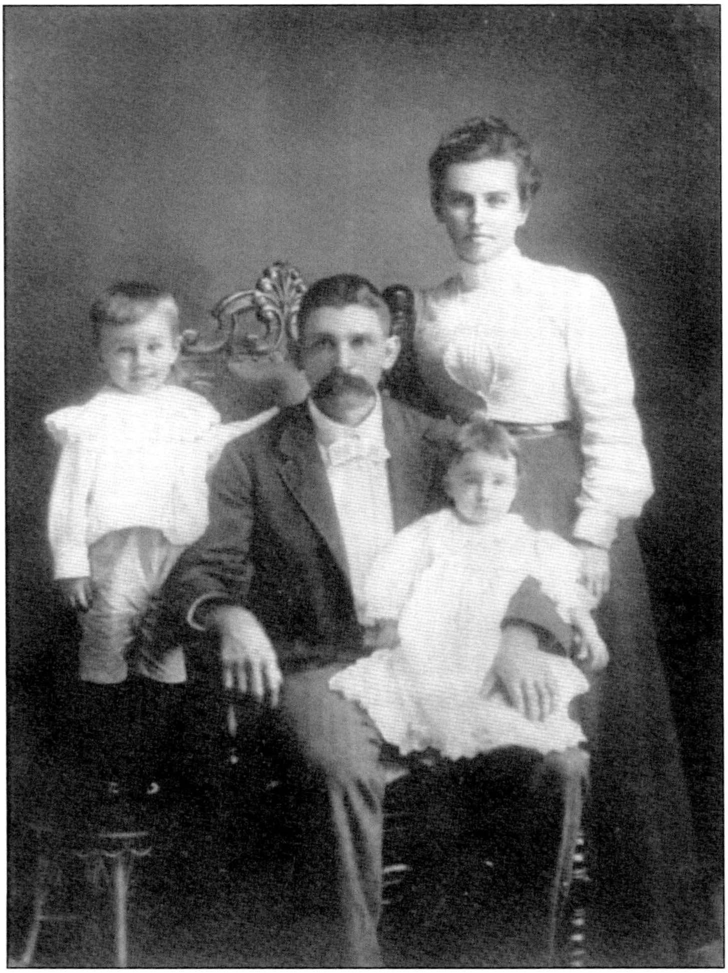

Sitting formally are Bertha Smoak Butler and Charlton Butler with children Newton (left) and Lena around the turn of the 20th century. Charlton was one of the more successful truck farmers in the parish and was instrumental in getting many of the major railroad lines and spurs laid throughout the parish. (Courtesy of Mona Butler Williams.)

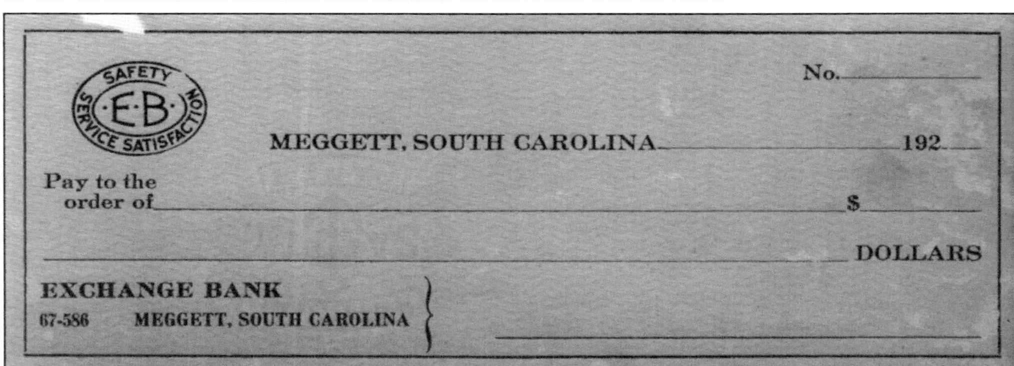

SAFETY · E·B · SERVICE SATISFACTION

No.______

MEGGETT, SOUTH CAROLINA______192__

Pay to the order of______ $______

______ DOLLARS

EXCHANGE BANK

67-586 MEGGETT, SOUTH CAROLINA

In 1912, Charles W. Geraty and other investors formed the Exchange Bank of Meggett for two reasons: there was no bank, and Geraty was looking for something to occupy his wife, Bessie, who at the time was grieving the loss of a child. The bank was housed on the ground floor of the Produce Association Building for over 30 years before merging with a larger bank chain in the 1950s. (Courtesy of William "Billy" Geraty.)

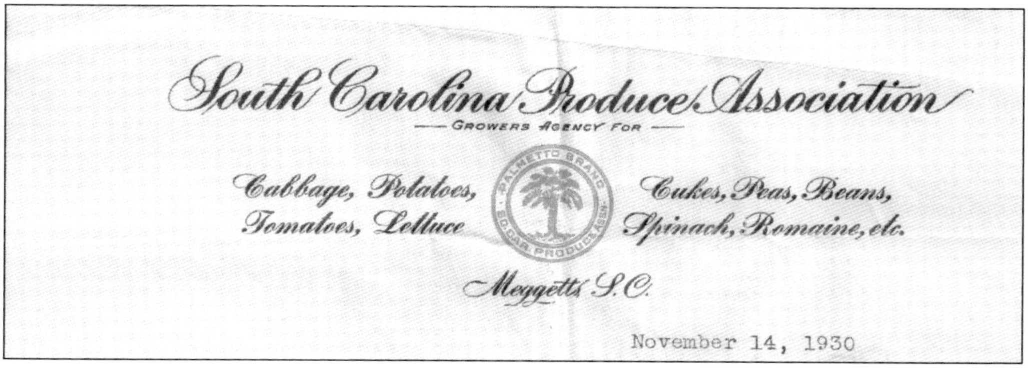
South Carolina Produce Association

— GROWERS AGENCY FOR —

Cabbage, Potatoes, Tomatoes, Lettuce

PALMETTO BRAND · SO. CAR. PRODUCE ASSN.

Cukes, Peas, Beans, Spinach, Romaine, etc.

Meggetts S.C.

November 14, 1930

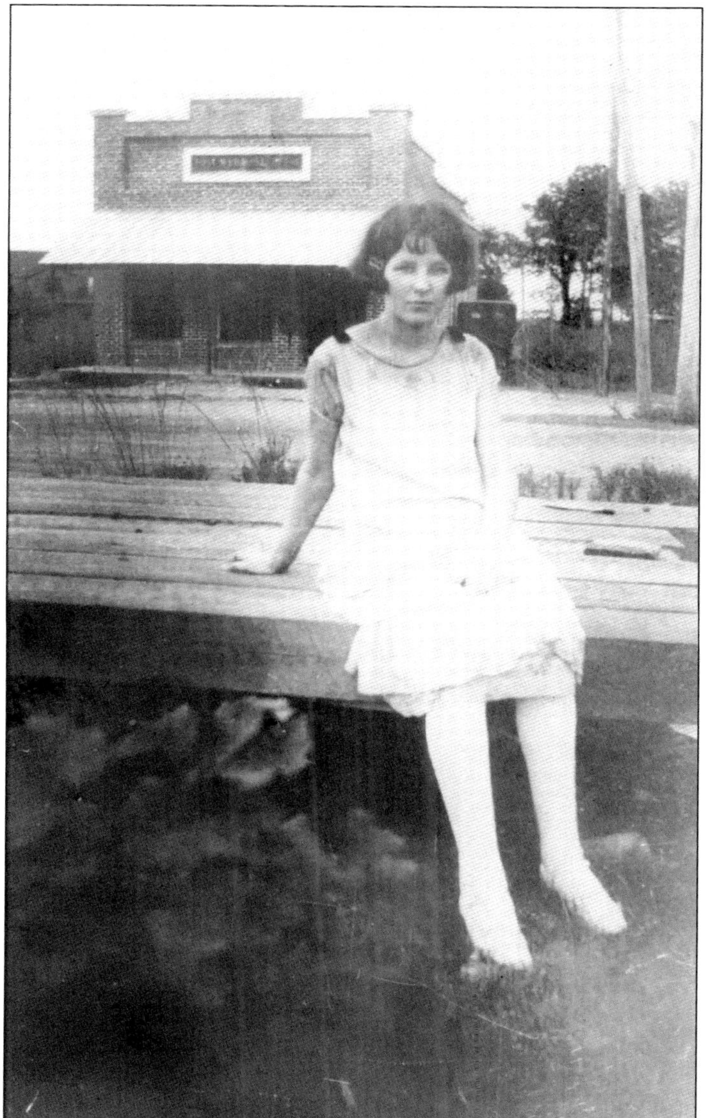

In the 1920s, several of the truck farmers in St. Paul's Parish formed two organizations—the South Carolina Produce Association, a strictly cooperative association, and the South Carolina Produce Exchange, a stock company. The association encouraged members to work together and assisted the farmers in marketing their products under the Palmetto brand. In addition, the farmers were able to collectively buy products like fertilizer and seeds in bulk, which was the most cost-effective way of doing business. At one point in time, the association became the largest distributor of vegetables in the Carolinas. The image above shows the letterhead that was used by the organization for all official correspondence. At right, Mary Dodd is sitting across the street from the two-story brick building that housed both the Exchange Bank and the offices for the Southern Produce Association. (Above, courtesy of Mona Butler Williams; right, courtesy of Anne Commins.)

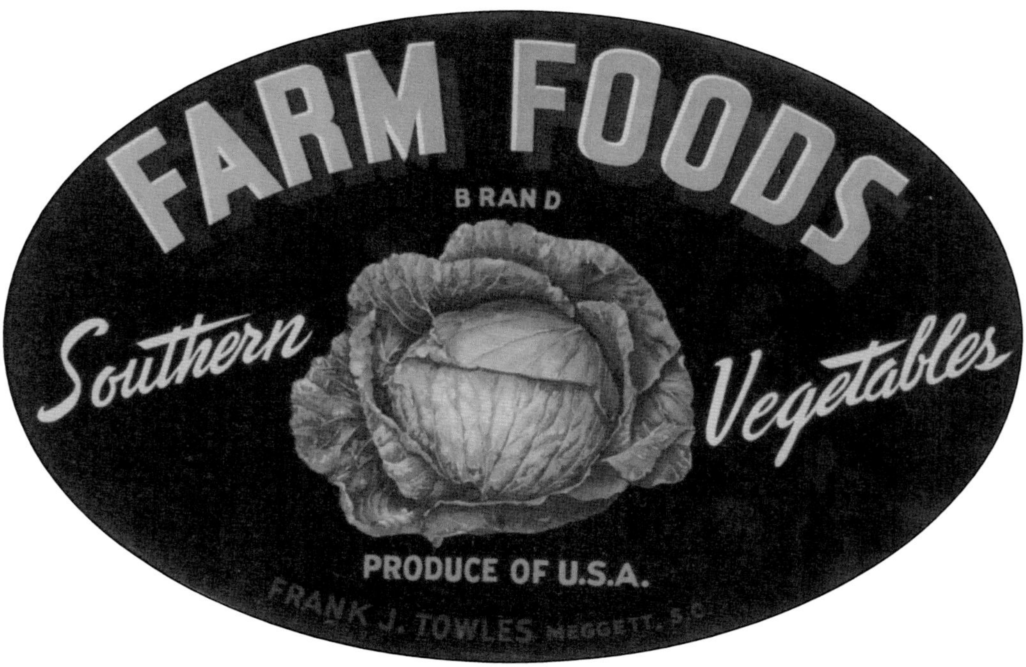

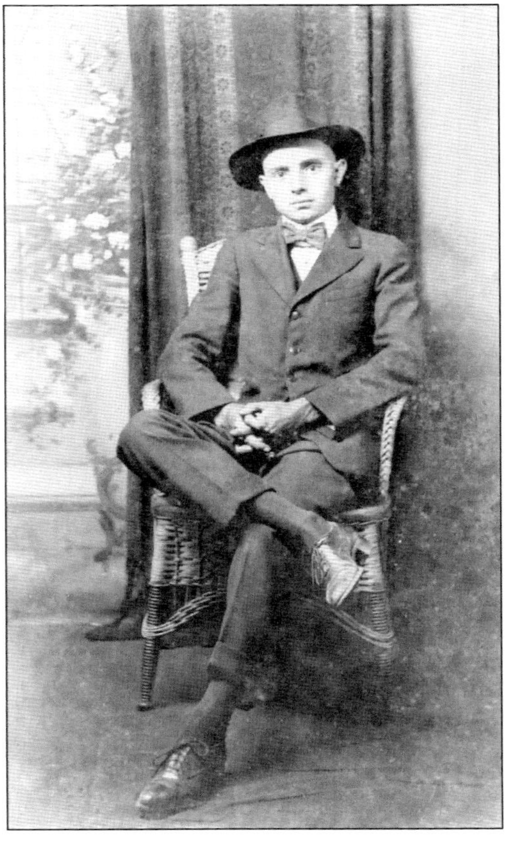

This sticker was designed and applied to an untold number of crates full of cabbages and other produce leaving the parish. During the 1912 harvest and shipping season, there were over 75 to 100 cars per day leaving the area. According to a *Charleston News and Courier* article, the total for the season included 4,000 carloads of cabbage, 1,000 carloads of potatoes, 600 carloads of beans, 800 carloads of cucumbers, and 500 carloads of miscellaneous crops. (Courtesy of John Waring and Andy Golden.)

Early on, Harry Gasque Boineau (1900–1949) was in charge of mail delivery out of the Osborn Post Office. He then went on to work as an automobile mechanic before ultimately ending up in charge of maintenance for Barrelville. John Mercer Heathington owned and operated Barrelville, which made the wooden crates and barrels that all the local produce was shipped in. (Courtesy of Jackie Boineau.)

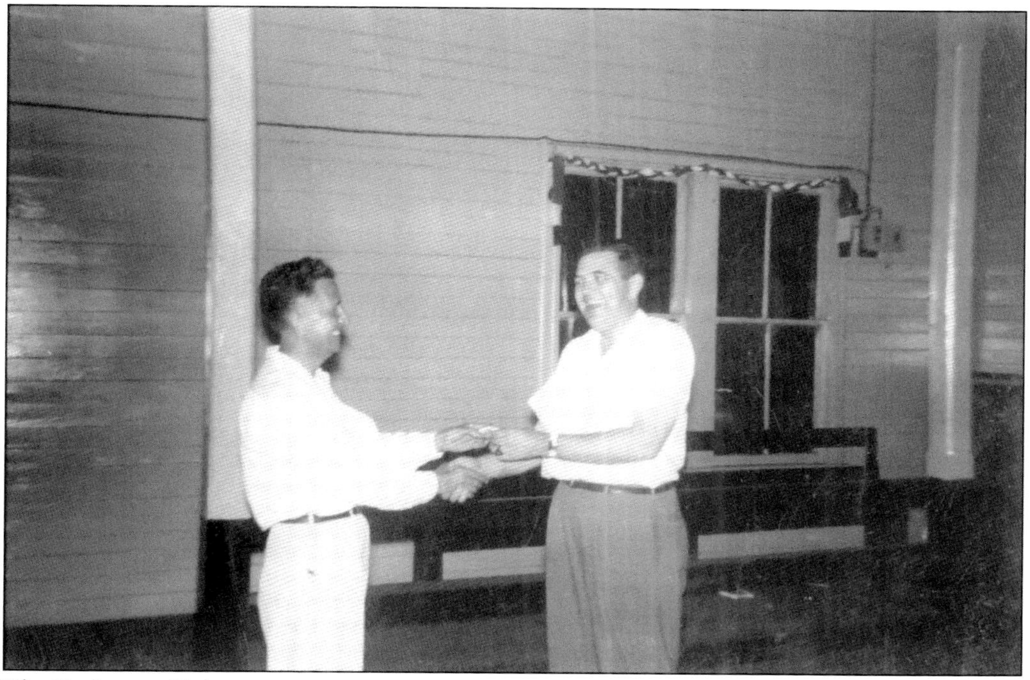

The Exchange Club was founded by Detroit businessman Charles A. Berkey in 1911. He chose the name "Exchange" because his idea was to gather like-minded individuals and businessmen to "exchange ideas and information" about how to better serve their communities. Above, Eddie Goodman (left) and Dr. Rogers are at a club meeting around 1950 in Hollywood, South Carolina. At right is a list of charter members of the Exchange Club of St. Paul's Parish. The membership is a who's who of businessmen and community leaders—and remains so today. (Above, courtesy of Anne Commins; right, courtesy of Mona Butler Williams.)

CHARTER MEMBERS

The Exchange Club of St. Paul's District (Meggett), S. C.

Barnwell, T. Osborne	Merchant
Bass, Rev. J. W.	Minister
Butler, Carl J.	Merchant
Butler, John M.	Farmer
Denham, Andrew	Farmer
Donaldson, Edward B.	Trucker
Donaldson, Pierce P.	Trucking
Ebner, J. W.	Merchant
Ebner, J. L.	Merchant
Fox, Innis L.	Barber
Fox, W. T.	General Store
Geraty, C. C.	Produce Comm.
Geraty, Charles W.	Farmer
Geraty, William C.	Farmer
Gibson, Charles E.	Vegetable Distbr.
Goodman, E. M.	Salesman
Griner, John F.	Auto Dealer
Ham, Arthur E.	Merchant
Henderson, Ralph L.	Lumber
Inabinett, A. C.	Grocer
Jenkins, E. Micah	Florist
Jenkins, John E.	Farmer
Johnson, John Q.	Furniture
Jones, E. K.	Farmer
Kelly, W. D.	Agent, A. C. L. R.R.
Kelly, B. B.	Lumber
King, Everett O. H.	Farmer
King, F. B.	Farmer
King, R. G.	Farmer
Kirver, Hoyt E.	Lumber
Lacy, M. L.	Lumber
Lacy, T. M.	Lumber
Lacy, W. A.	Lumber
Lee, F. F.	Produce Sales
Lomax, L. M.	Supt. Lumber Co.
Martin, Archie H.	Wood Work
Matarese, George J.	Restaurant
McLeod, M. L.	Lumbering
Moody, D. F.	Mail Carrier
Murray, J. G.	Produce Sales
Newton, W. J.	Lumber
Newton, Leonard E.	Saw Mill Opr.
Parish, George A.	Farmer
Pope, Jenkins M., Jr.	State Park Supt.
Prine, Joseph L.	Ice Industry
Puckett, Guy C.	Service Manager
Reeves, David C.	Merchant
Roberts, James H.	School Principal
Rogers, Dr. Robert S.	Physician
Sanders, Edward C.	Clerical
Sanders, John S.	Auto Dealer
Sandifer, J. B.	Parts Manager
Schroeder, R. D.	Superintendent
Searson, W. B.	Merchant
Sires, Norman L.	Service Station
Stevens, S. Y.	Marine Line Co.
Stevens, W. J.	River Transp.
Thompson, Julius P.	Trucking
Towles, Joe F.	Produce
Towles, Frank	Farmer
Weikert, Bernard J.	Bookkeeper
Whaley, W. J.	Merchant
Whetstone, M. J.	Theatre
Ypuman, A. B.	Gas Distributor

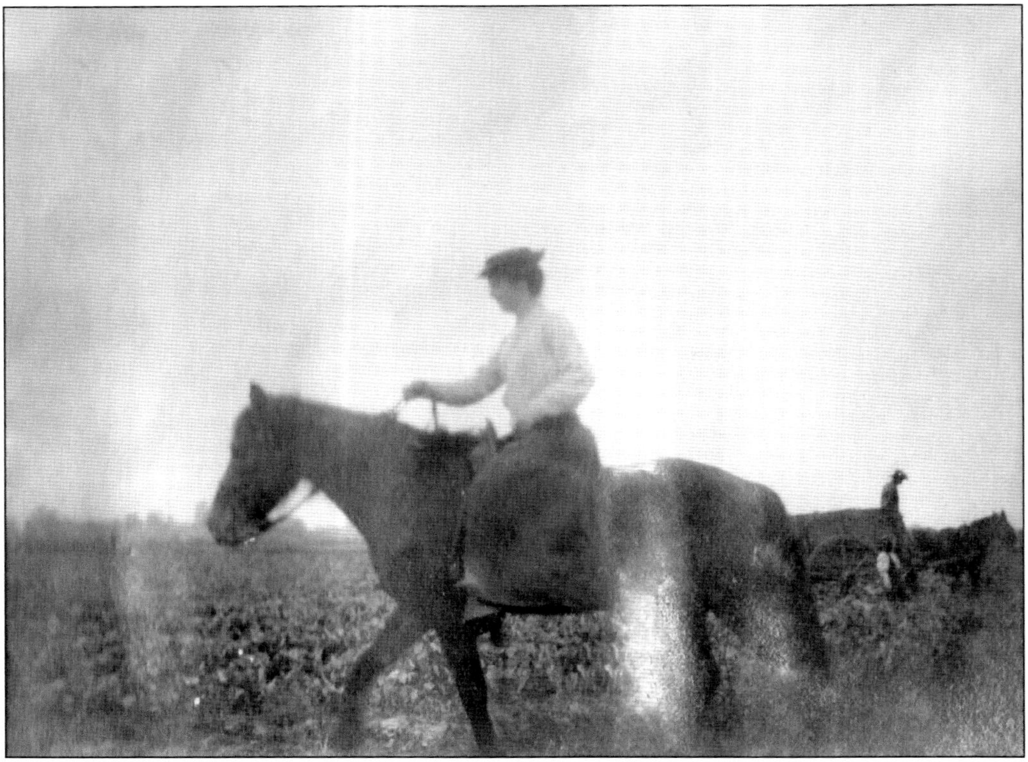

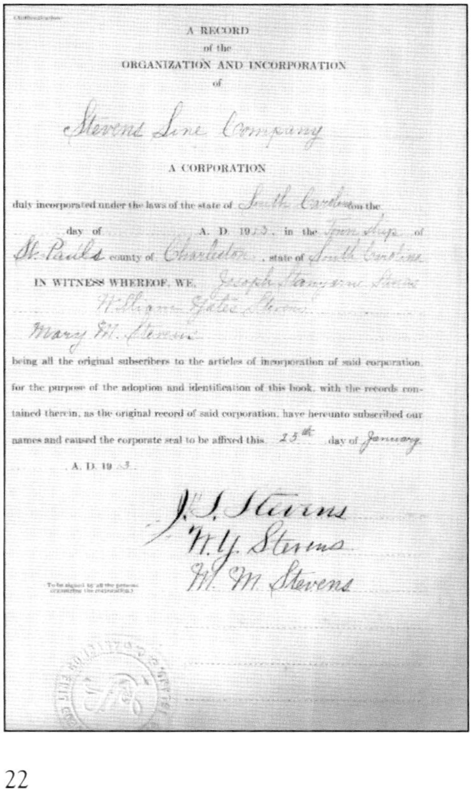

A RECORD
of the
ORGANIZATION AND INCORPORATION
of

Stevens Line Company

A CORPORATION

duly incorporated under the laws of the state of South Carolina on the ___ day of ___ A. D. 1913, in the Township of St. Pauls county of Charleston, state of South Carolina

IN WITNESS WHEREOF, WE, Joseph Stanyarne Stevens, William Yates Stevens, Mary M. Stevens

being all the original subscribers to the articles of incorporation of said corporation, for the purpose of the adoption and identification of this book, with the records contained therein, as the original record of said corporation, have hereunto subscribed our names and caused the corporate seal to be affixed this 23th day of January A. D. 1913.

J. S. Stevens
W. Y. Stevens
M. M. Stevens

Here, Hilmer "Hillie" Behling rides her horse Bonnie Jean on Rescue Plantation in the spring of 1911. Farm animals, in particular horses, were considered a valuable asset before the dawn of tractors and trucks. (Courtesy of Sarah Roberts.)

Until the 1930s, there were few roads and bridges on the South Carolina coast, and boats were the only viable form of transportation in rural areas. Large steamers ran regular schedules to and from towns, carrying people, crops, and mail. The Stevens Line of Yonges Island was one of several family enterprises that hauled freight and carried mail. The Stevens Line Company began in the 1880s and was incorporated, as the document at left indicates, in 1913. In the late 1970s, the company became Stevens Towing Co. Inc., and it is still operating today. (Courtesy of Susan Stevens.)

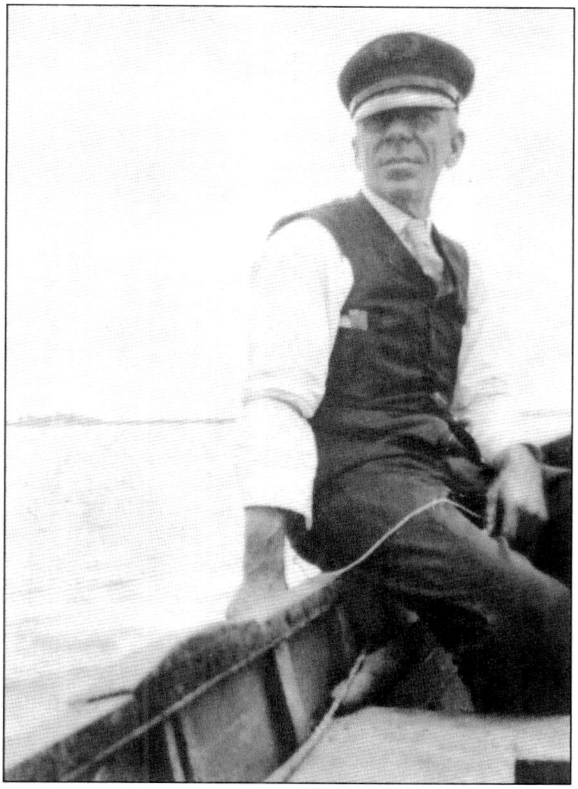

At right is Capt. J.S. Stevens in the early 1900s. J.S. and W.Y. Stevens began the line in the 1880s when they built their first boat, *Otto*. At its peak in the 1930s, the Stevens Line had from 10 to 20 boats operating, as well as several barges. Below is the *Sunbeam*, one of the Stevens boats used to move produce and goods. Oftentimes, these were called "cabbage boats" for the large amounts of cabbages they transported to market for the local farmers. (Right, courtesy of Susan Stevens; below, courtesy of Susan Stevens and John Waring.)

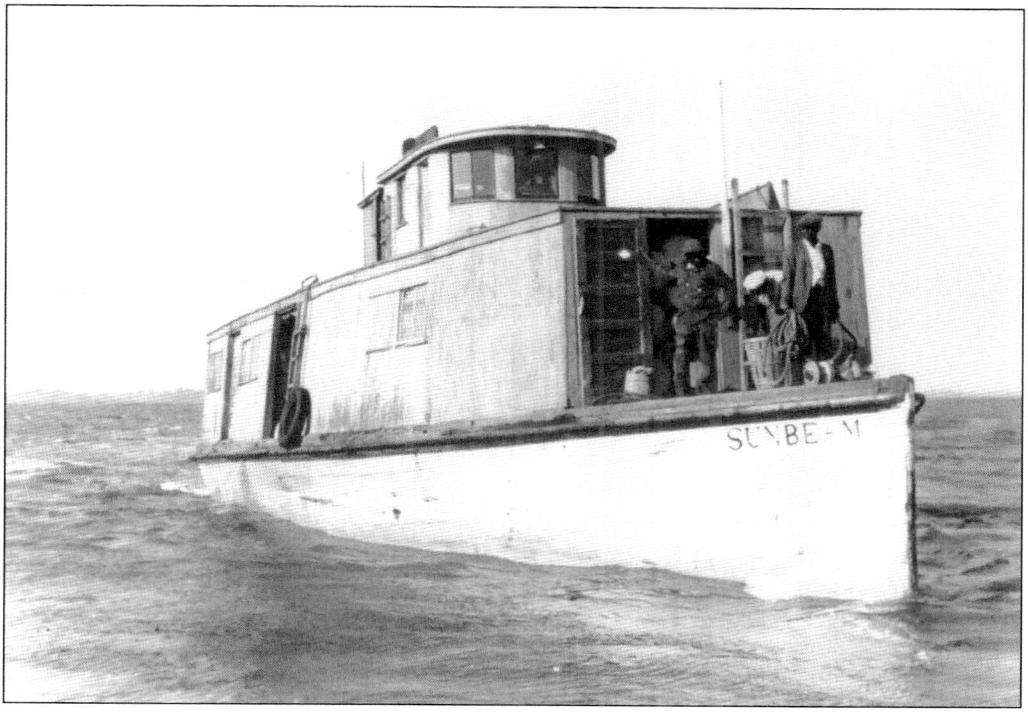

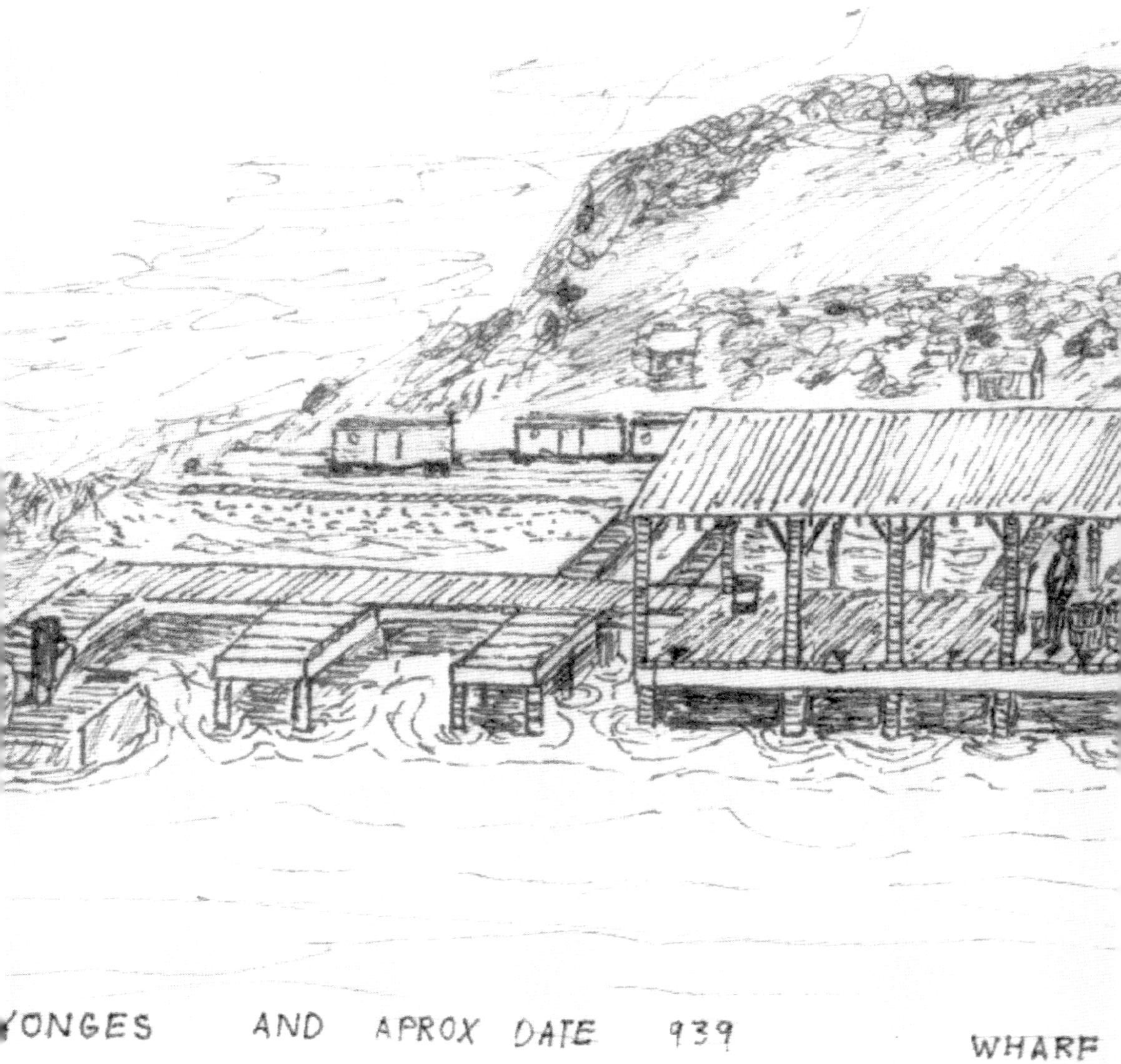

Yonges Island was part of a 1694 royal land grant of 640 acres to Thomas Smith, who paid one penny per acre "in fine silver." In the latter half of the 18th century, it was owned by James Gilchrist Simpson and his wife. Simpson sold the northern side of the island to Isaac Lesesne in 1771 and the southern side to Francis Yonge around 1775. At an unknown later date, Yonge

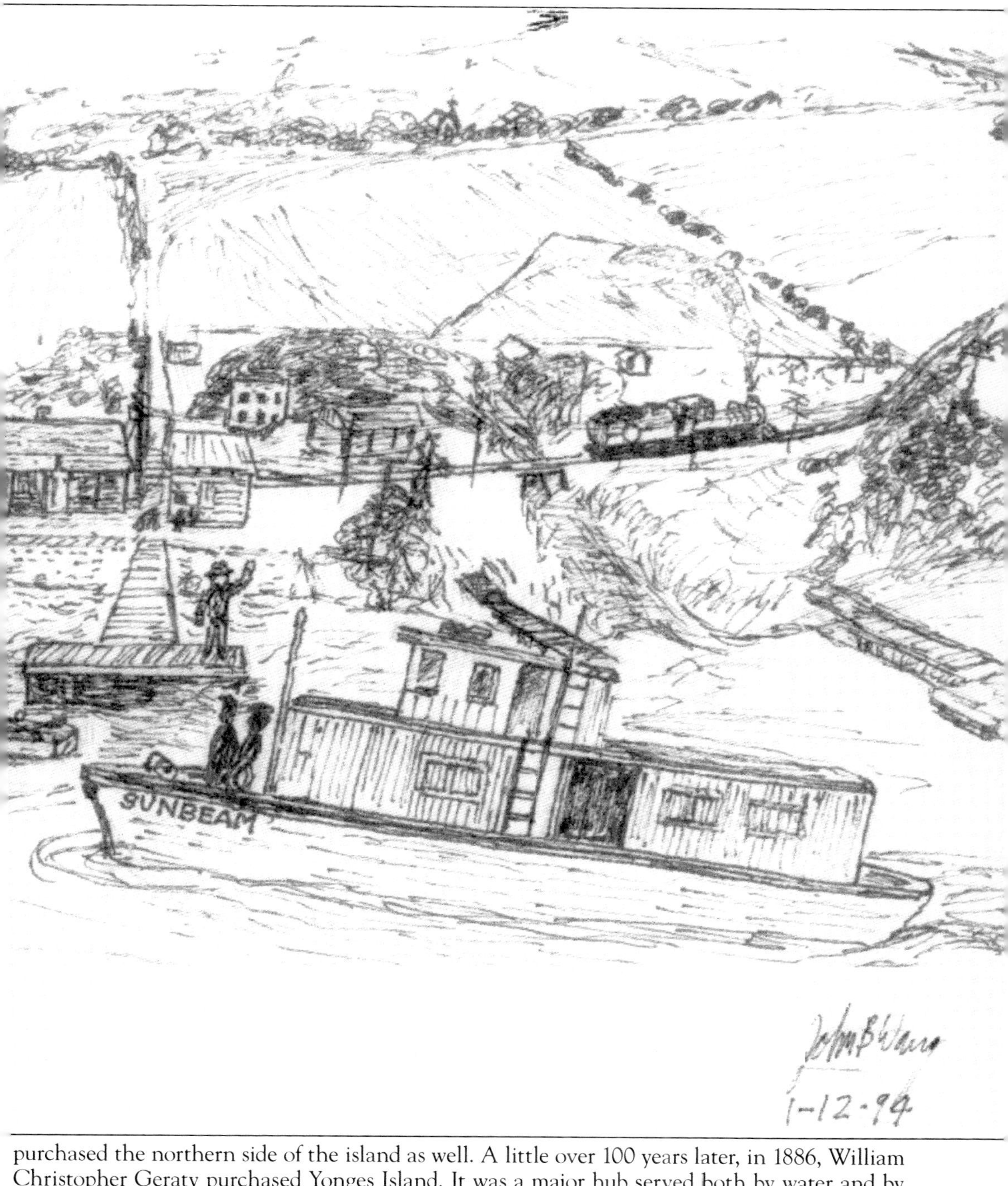

purchased the northern side of the island as well. A little over 100 years later, in 1886, William Christopher Geraty purchased Yonges Island. It was a major hub served both by water and by rail. Seen here is a sketch created by local artist John Waring of the wharf at Yonges Island. (Courtesy of John Waring.)

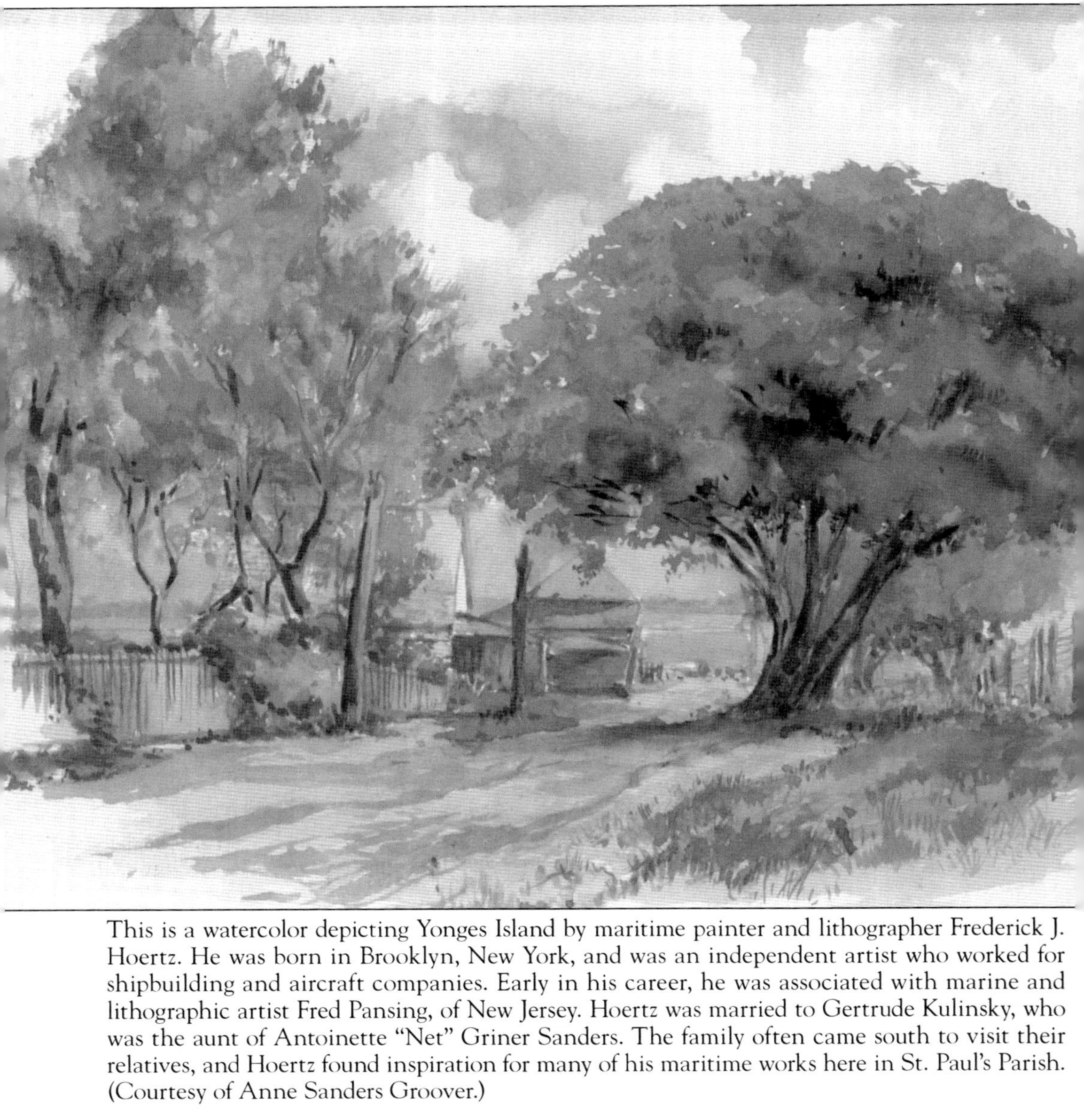

This is a watercolor depicting Yonges Island by maritime painter and lithographer Frederick J. Hoertz. He was born in Brooklyn, New York, and was an independent artist who worked for shipbuilding and aircraft companies. Early in his career, he was associated with marine and lithographic artist Fred Pansing, of New Jersey. Hoertz was married to Gertrude Kulinsky, who was the aunt of Antoinette "Net" Griner Sanders. The family often came south to visit their relatives, and Hoertz found inspiration for many of his maritime works here in St. Paul's Parish. (Courtesy of Anne Sanders Groover.)

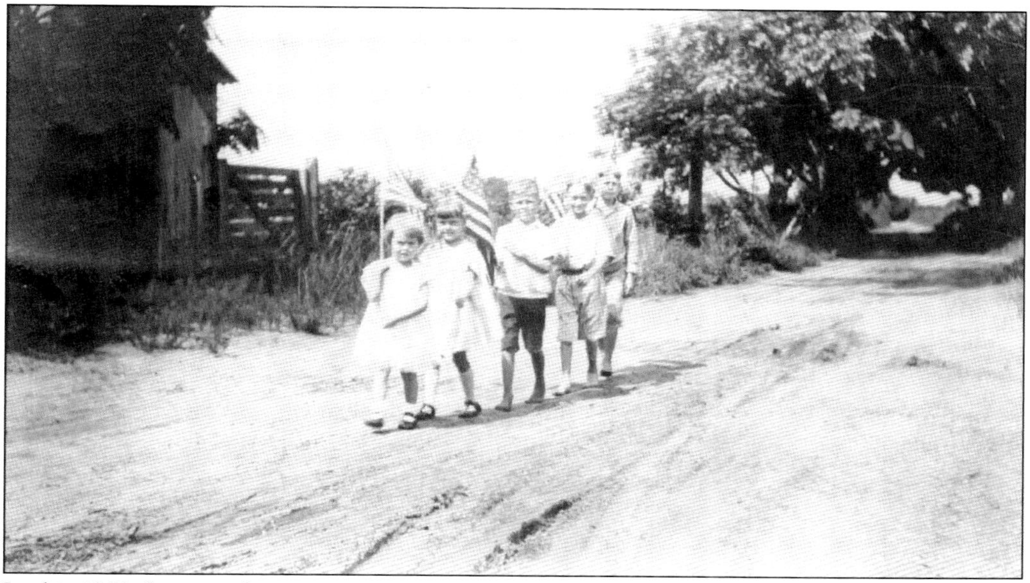

In this 1917 photograph, from left to right, Charlotte Hoertz, Net Griner, John Griner, Edwin Hoertz, and Edward Harter are walking down a dirt road waving American flags to celebrate the Fourth of July in a spontaneous parade. (Courtesy of Anne Sanders Groover.)

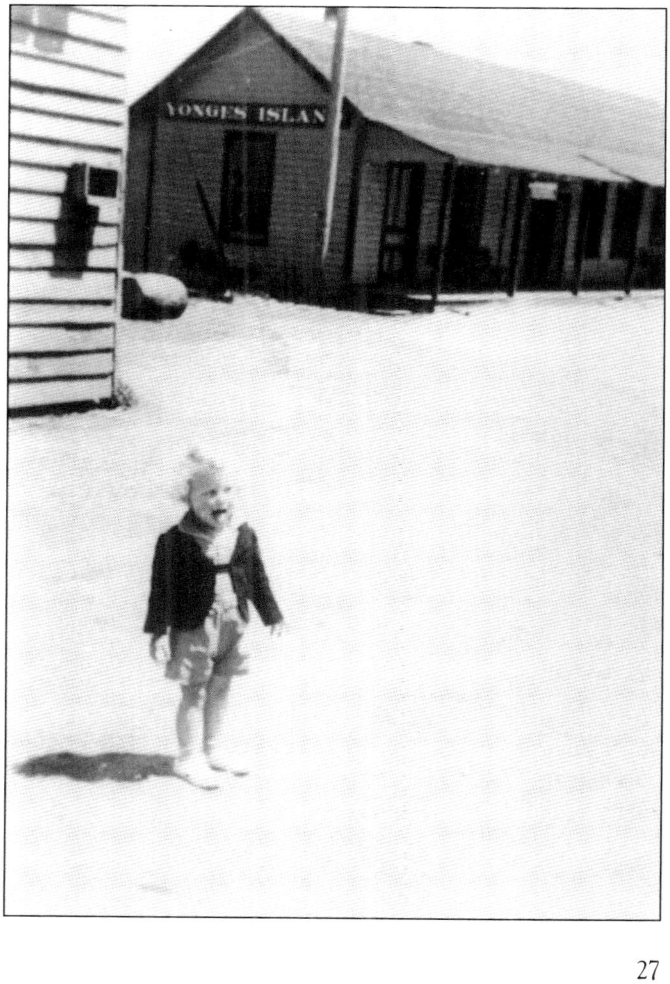

Pictured around 1920 is an unidentified child standing in front of the Yonges Island railroad depot, which was a main transportation hub for the farmers and fisherman of the island. (Courtesy of John Waring.)

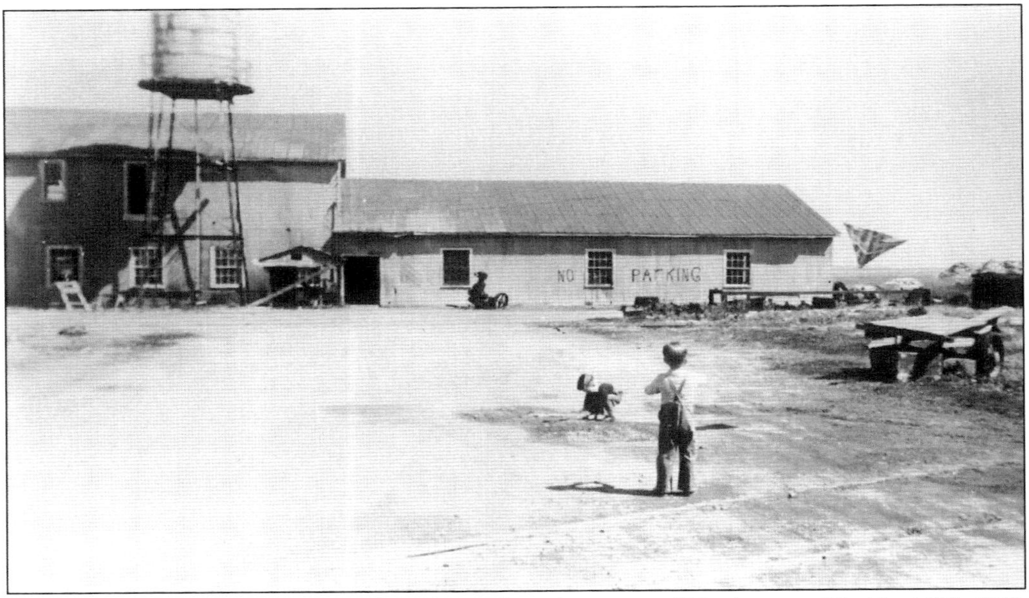

According to *Tidecraft: The Boats of South Carolina, Georgia, and Northeastern Florida, 1550–1950,* L.P. Maggioni, a native of Monza, Italy, founded a wholesale operation at Savannah in the 1800s supplying fish and oysters from local waters. The Maggioni Company eventually grew into the largest canner and supplier of seafood in the Southeast for much of the 1900s. Together with sister companies, Maggioni owned 130 boats and five canneries that could process up to 7,000 bushels of oysters per day. Seen here are Jimmy Powers and another child flying a kite outside of the Yonges Island Oyster Factory. Jimmy's father assisted the Maggioni family in running the oyster factory on the island. (Courtesy of John Waring.)

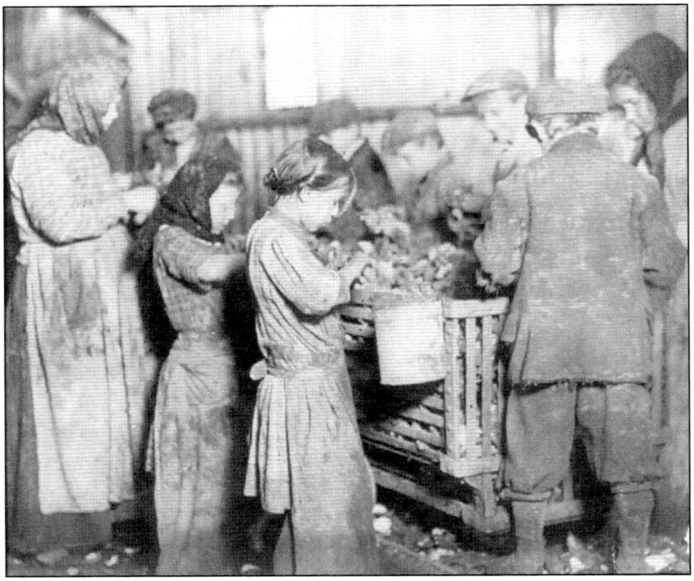

Several women and children, mostly between the ages of 10 and 12, are shucking oysters in 1913 in the Varn and Platt Canning Company on Yonges Island in a photograph taken by Lewis Wickes Hines. Hines began taking pictures for the National Child Labor Committee in 1908, and many of his images of children working on farms and factories helped to change child labor laws in the United States. (Courtesy of the Lewis Hines Collection at the Library of Congress, Washington, DC.)

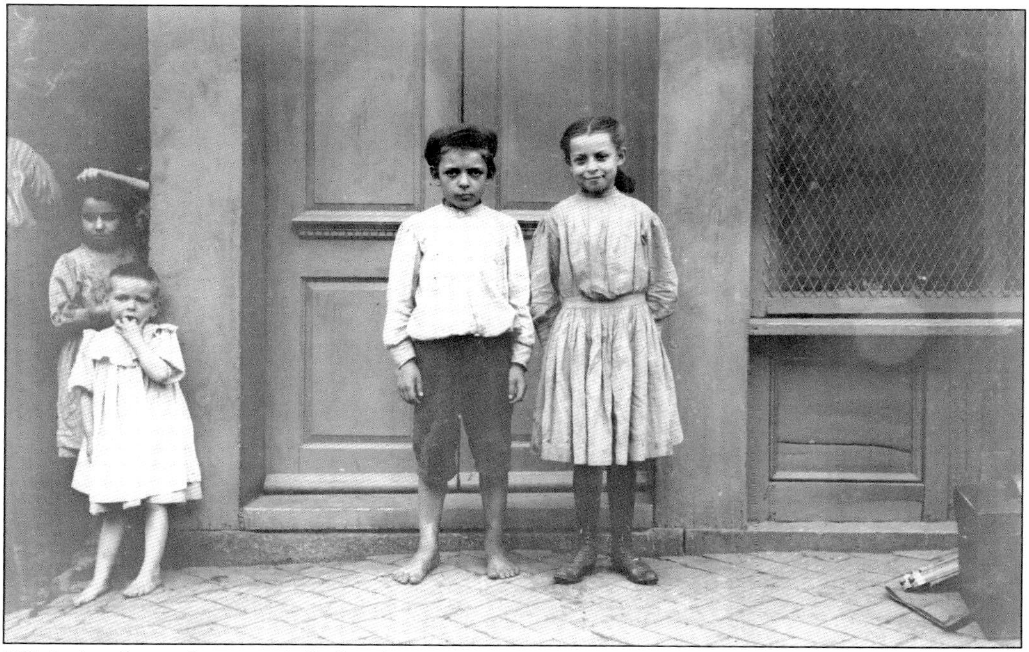

While local people were skilled at harvesting and shucking oysters, they were unfamiliar with industrial canning operations. Therefore, most of the experienced cannery workers were found up North and brought down South to work in the factories; Yonges Island was no exception. Most of the workers were of Polish descent and came down each winter from cities such as New York, Philadelphia, and Baltimore to labor in the factories. Many times, they brought their children with them, and the children were also pressed into service. Here, Marie and Albert Kawalski (center) are standing at 615 Band Street in Baltimore, Maryland. Marie is 11, and Albert is 10. They both had worked with their mother the previous winter shucking oysters on Yonges Island. Below, four-year-old Mary (foreground) is helping other shuckers inside the Varn and Platt Canning Company on Yonges Island. (Both, courtesy of the Lewis Hine Collection, Library of Congress, Washington, DC.)

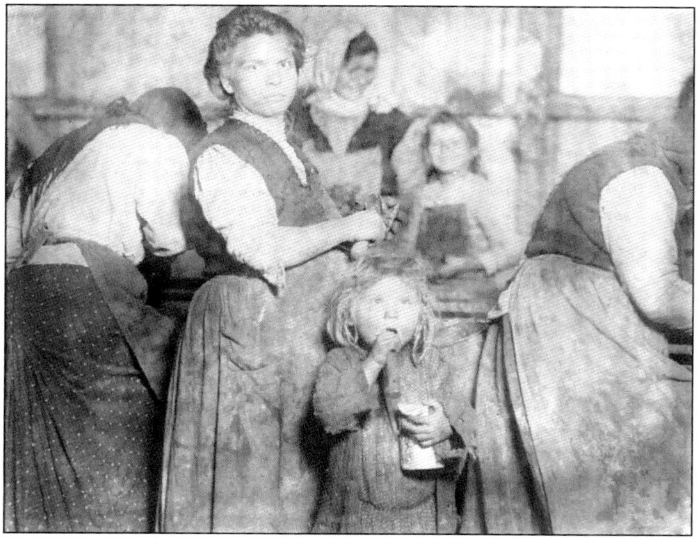

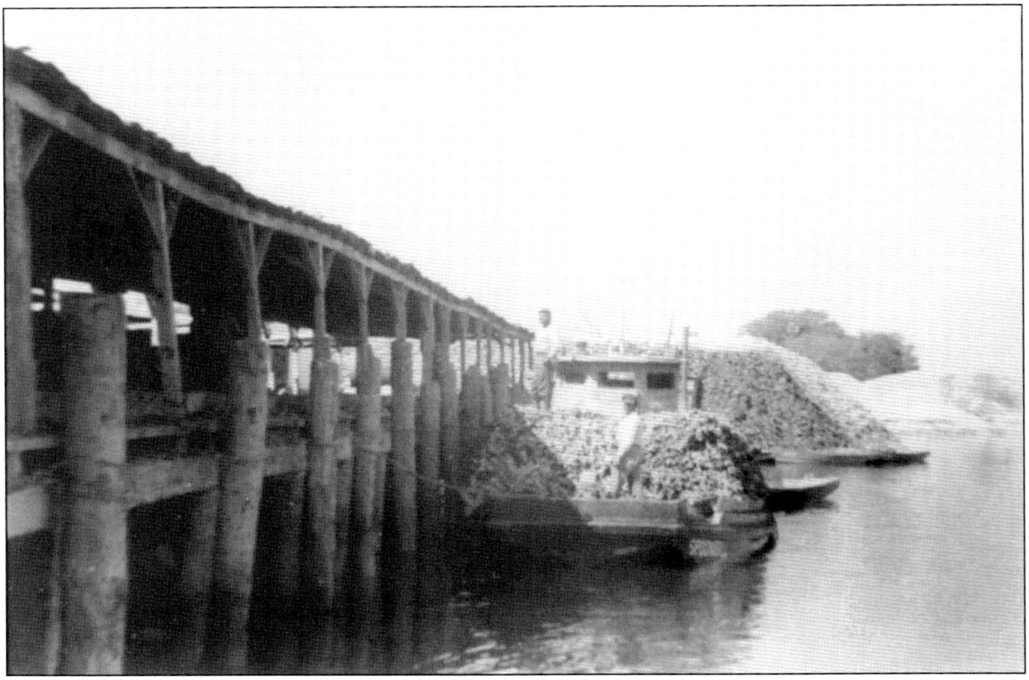

Pictured is the dock outside of the oyster factory where oyster boats would off-load their catches to be processed. The date of the picture is not known but believed to be in the early 1940s. Below is the same dock after a Category 2 hurricane struck South Carolina in the fall of 1945. It is believed the dock was completely destroyed in 1954 when Hurricane Hazel hit the area. (Both, courtesy of John Waring.)

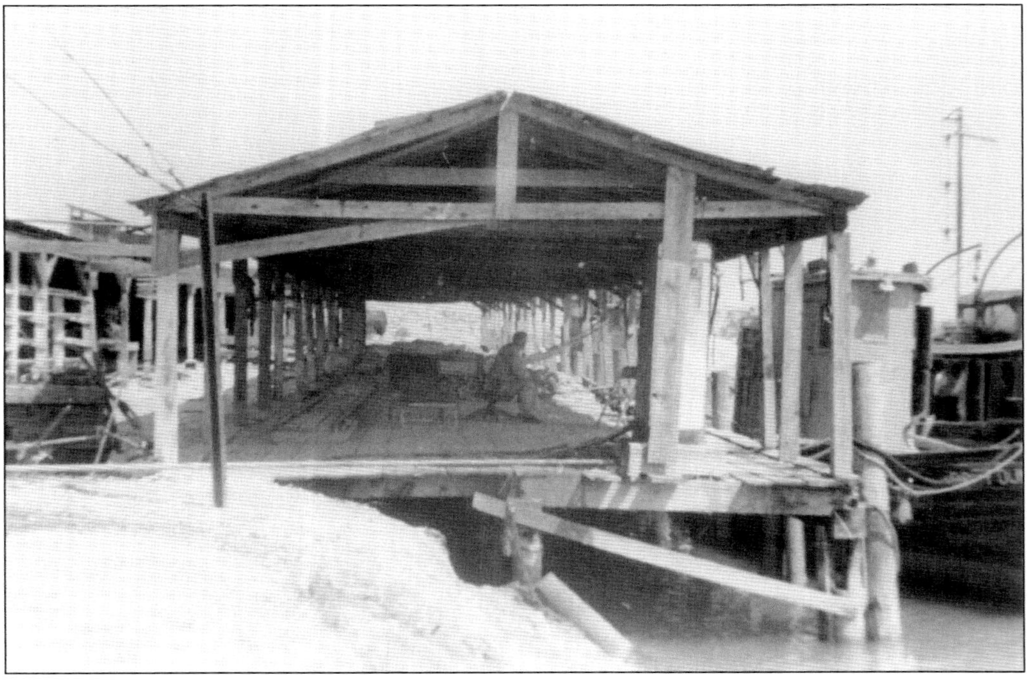

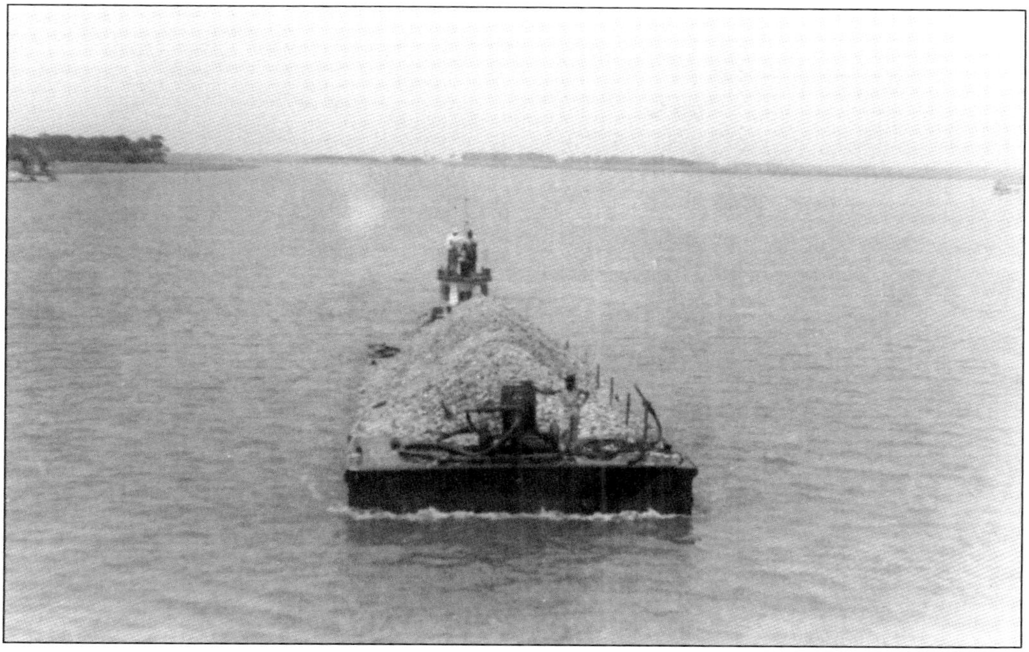

According to an article titled "The Salty Dogs" in the Winter 2000 issue of *Coastal Heritage* magazine, during the fall and winter, local fisherman would go out on flat-bottomed boats called bateaux or in larger sloops or engine-powered scows when they would harvest oysters from the riverbanks. It was thought that workers could gather anywhere from 60 to 100 bushels during one low tide. The oysters were then transported to factories, such as the one on Yonges Island, where they were steamed, shucked, and canned. Above are the images of two such oyster boats that operated off Yonges Island. (Both, courtesy of John Waring.)

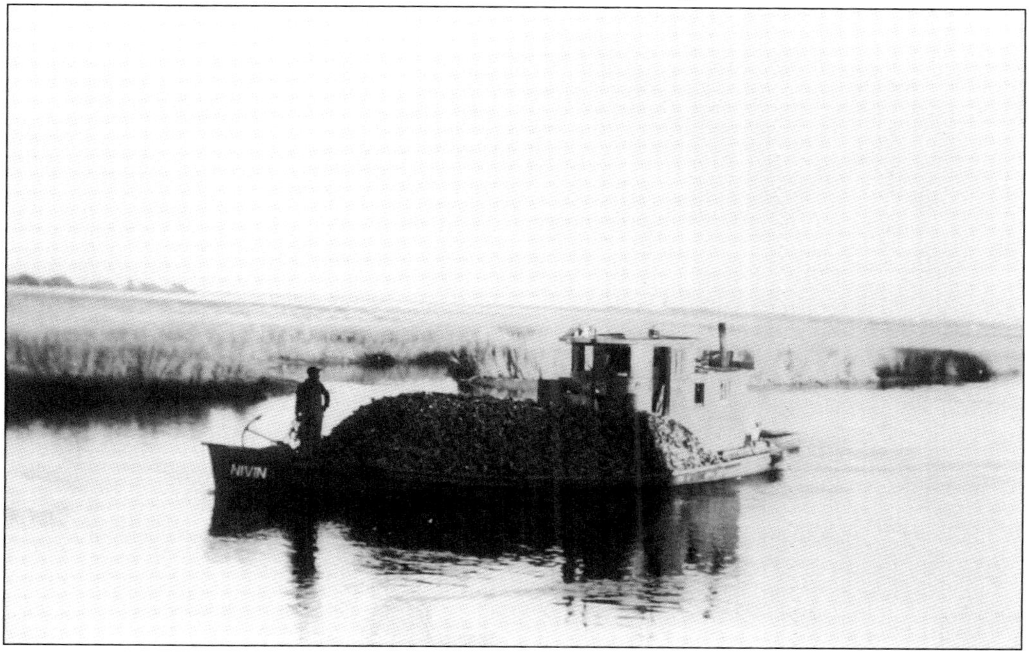

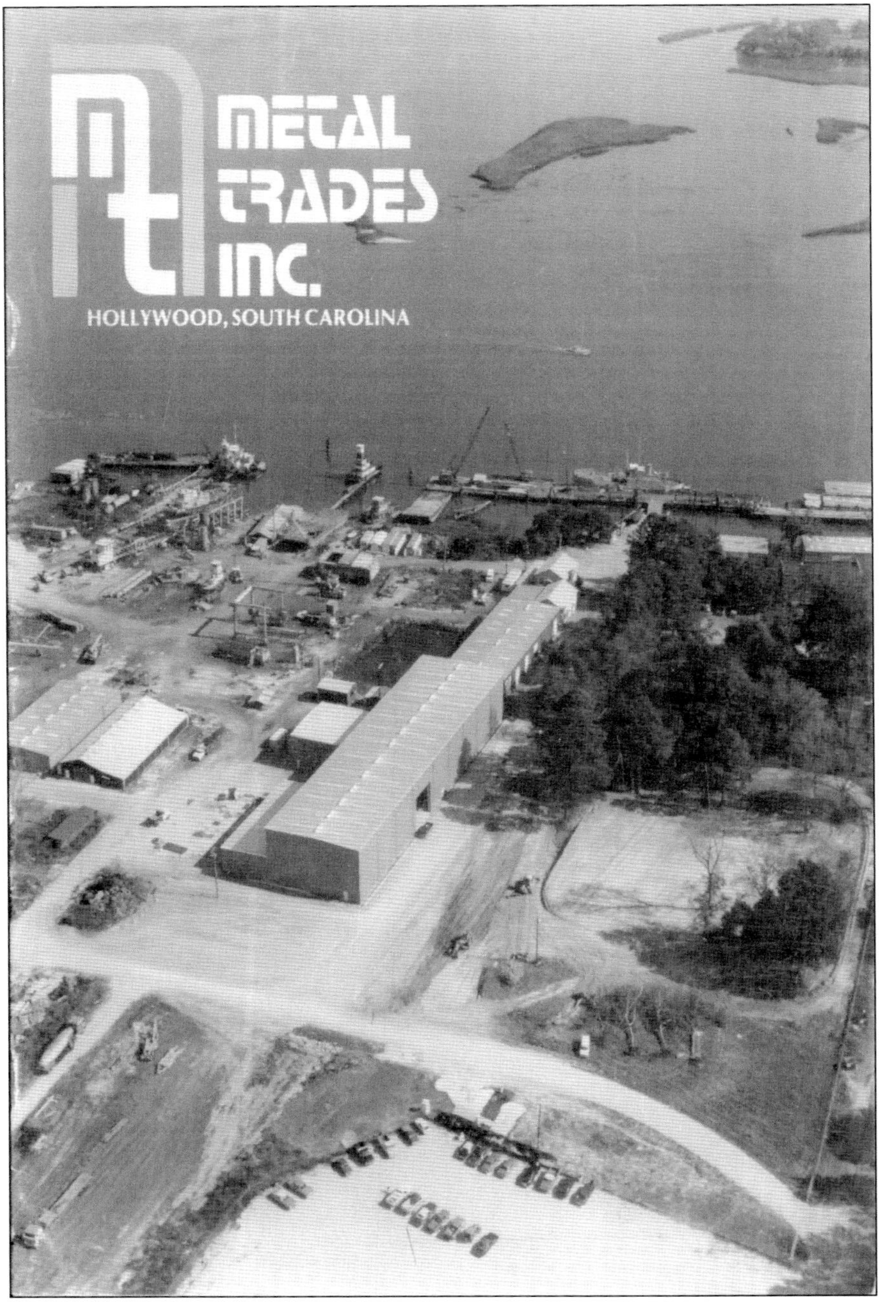

J.E. Corbin founded Metal Trades Inc. (MTI) in 1962 with a pickup truck, a welding machine, and years of metal fabricating experience. In the early years, MTI focused on steel and sheet metal fabrication. The company later became certified as a master ship-repair contractor for the US Army and Navy, providing overhaul and ship-repair services on various types of coastal vessels. MTI established a solid reputation for its multiskilled workers and strong commitment to high quality. In recent years, MTI has demonstrated proven expertise in barge construction. The company continues to be family-owned-and-operated. (Courtesy of Metal Trades Inc.)

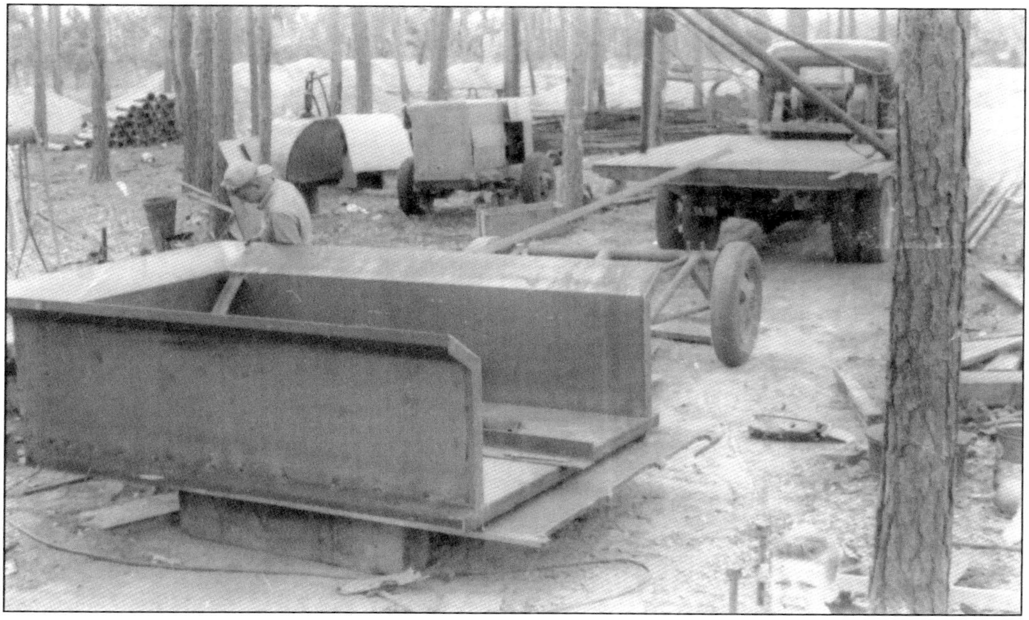

Seen above around 1952 with four-year-old Eddie Corbin (in the right foreground) is the beginning of what was to become a major family-owned corporation based out of Yonges Island. Below is Joe Corbin, founder of Metal Trades Inc., captured walking around the yard inspecting the work being done by Metal Trades workers in his later years. (Both, courtesy of Metal Trades Inc.)

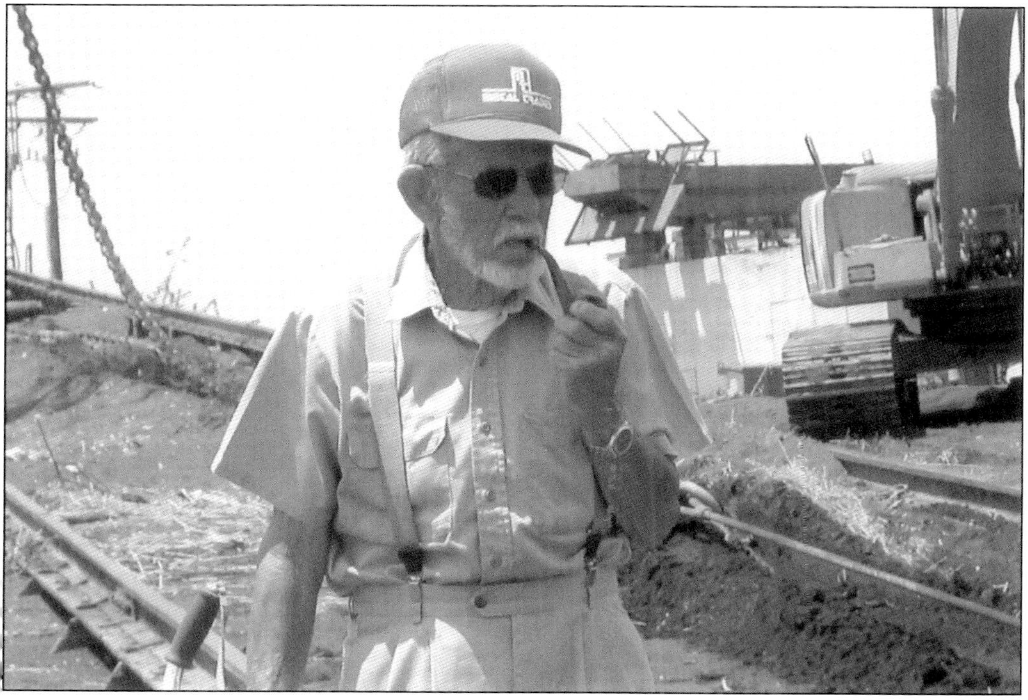

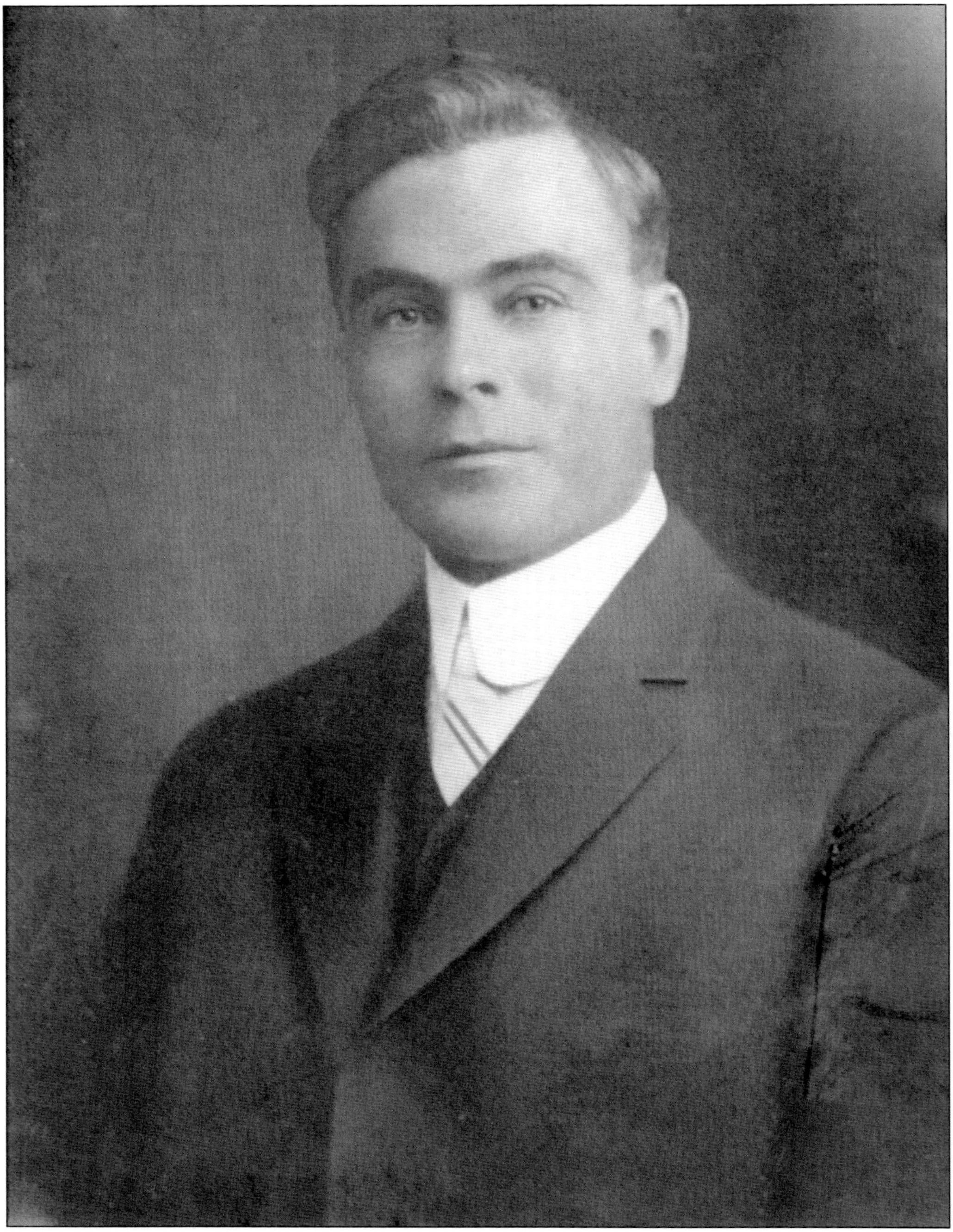

The Towles family has deep roots in St. Paul's Parish and began farming in the South Carolina Lowcountry in the mid- to late 1800s on Wadmalaw Island, but they moved much of their farming operations to Yonges Island and Meggett around the turn of the century. Francis J. Towles (pictured), born April 7, 1908, was the son of Frank E. Towles and Martha Sara Wilson. (Courtesy of Beth Van Norte.)

Two

GATHERING PLACES

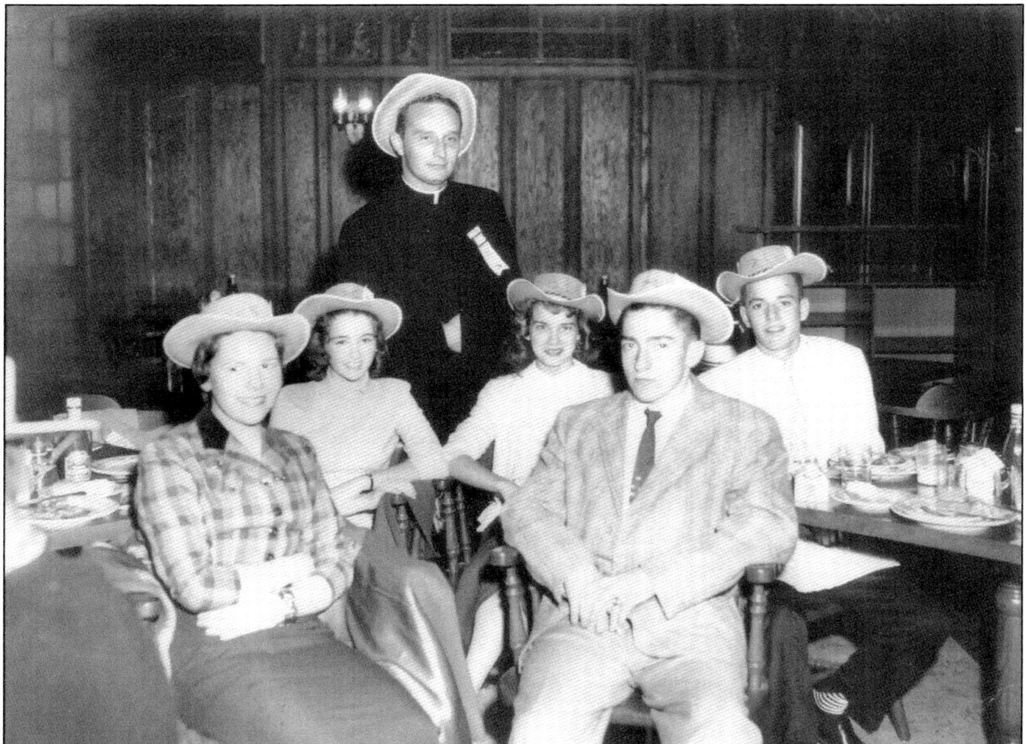

Whether at the local country store, gas station, church, school, outside, or at someone's home, people gather to share stories, celebrate birthdays, remember loved ones, or simply to discuss their children, what is going on in town, and to catch up on the latest gossip. In the photograph are, from left to right, Anne Sanders, Cindy Harrison, Fr. Fleming McManus, Mary Agnes DeChristie, Bubba Moody, and Nicky Butler. The picture was taken on a youth trip to Washington, DC, in the 1960s. (Courtesy of St. Mary's Catholic Church.)

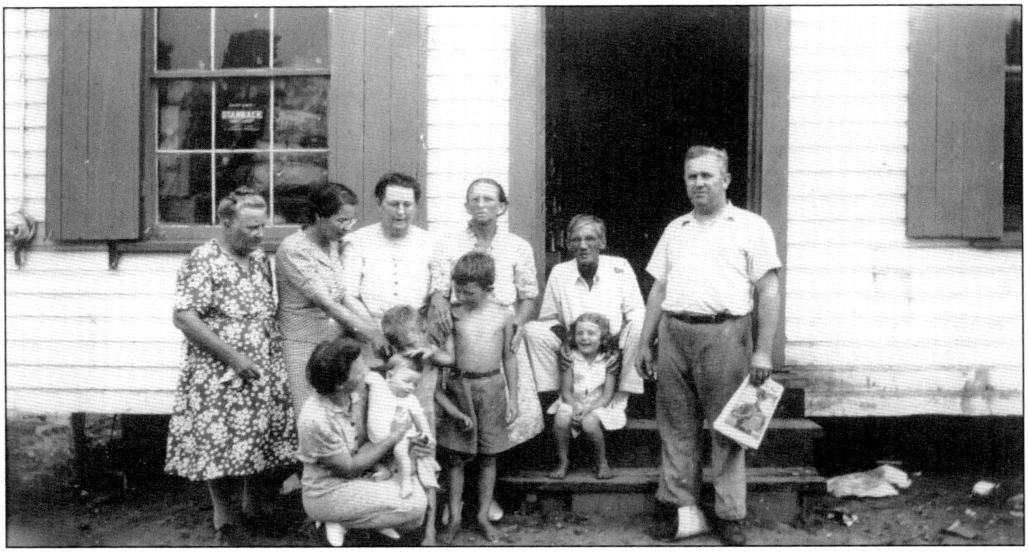

These are members of the Bair family relaxing outside of the Bair's Store, which was located at the corner of Towles Road and Highway 162 outside of Hollywood. The building is still standing today, though in a state of disrepair. Family-owned shops were generally located about one mile apart so that residents could walk to the store to buy necessary groceries and other household items. (Courtesy of Judy and Reeves Bair.)

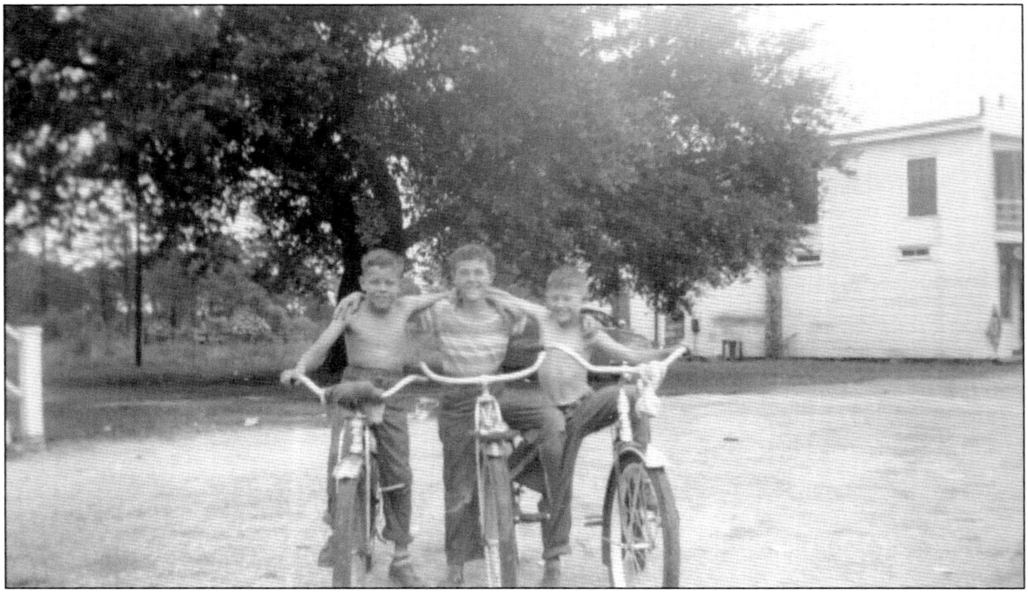

In this photograph taken in 1952 in downtown Hollywood are three local boys, all age 10, riding their bikes. From left to right are Ron "Ronnie" Rash, David Reaves Jr., and Henry Thompson. In the background is Jenkins Grocery Store. (Courtesy of Ron Rash.)

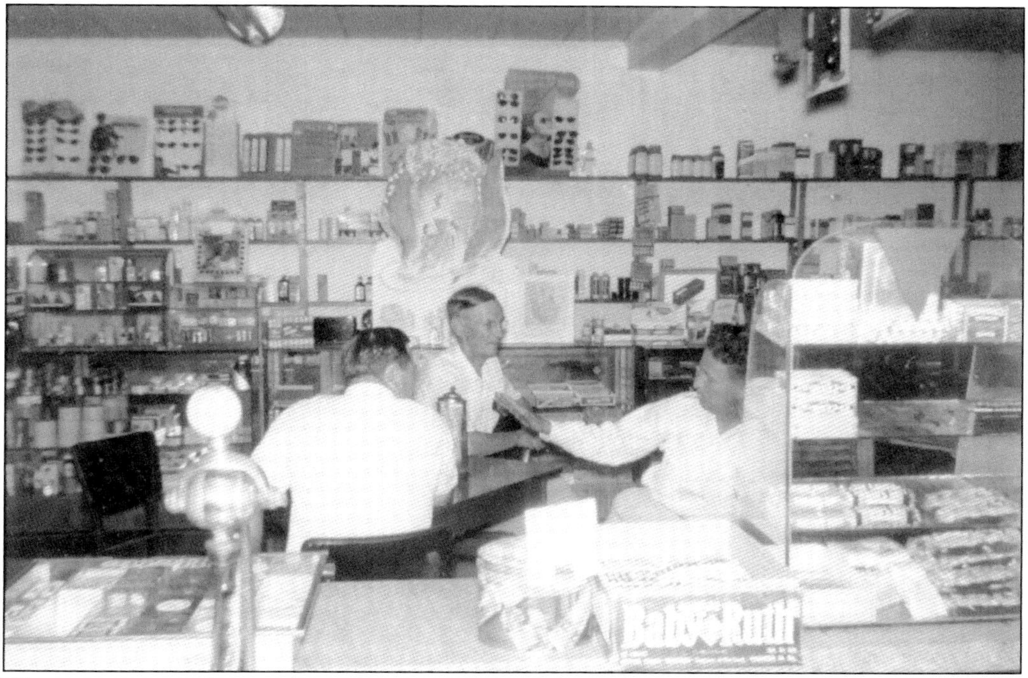

This picture was taken in the early 1950s inside of the Hollywood Fountain, the local drugstore and soda fountain shop on Highway 162 in Hollywood. Eddie Goodman (right), a salesman for Griner Chevrolet, and two unidentified men are having lunch. David Reaves operated the store from September 1943 until 1952. (Courtesy of Anne Commins.)

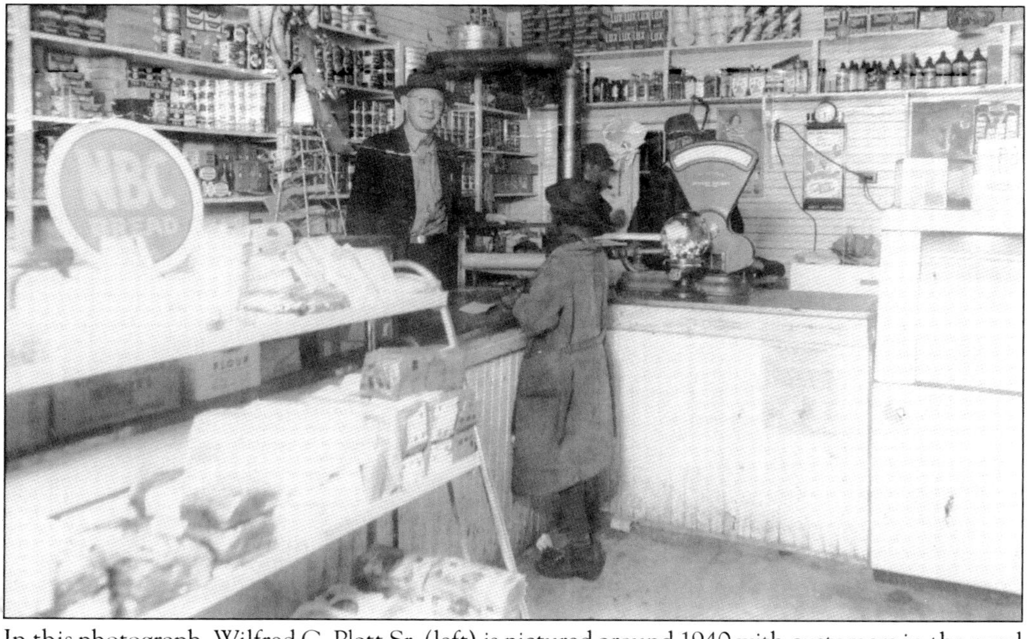

In this photograph, Wilfred C. Platt Sr. (left) is pictured around 1940 with customers in the rural grocery store that he and his wife, Lucille Weeks Platt, built and operated on Highway 162 between Hollywood and Adams Run, South Carolina. (Courtesy of Elizabeth "Betsy" Waters.)

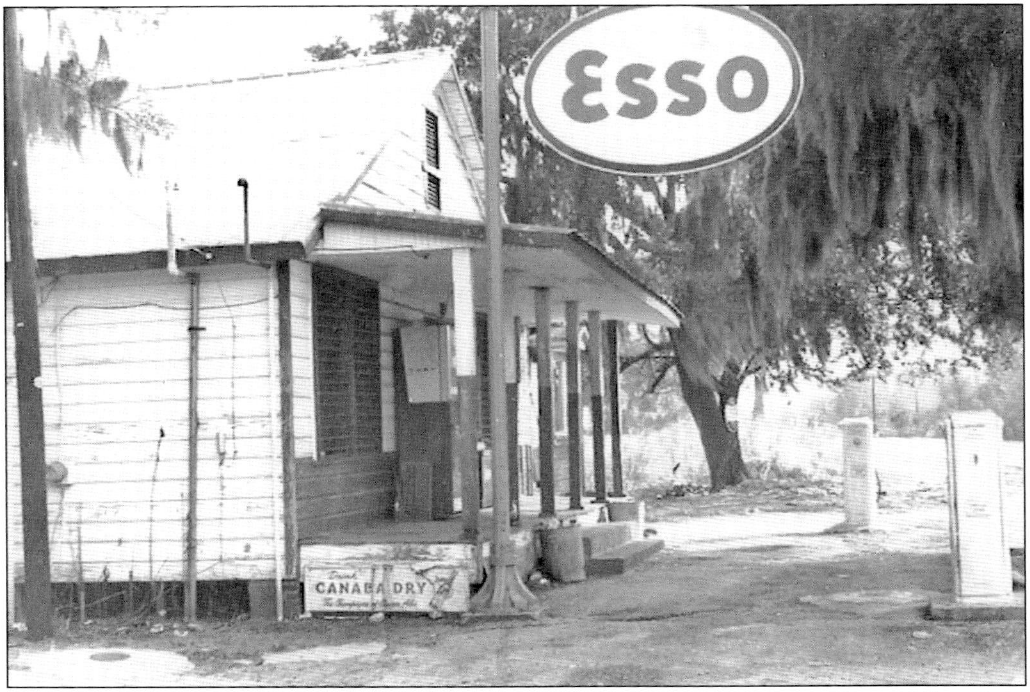

Roxbury Mercantile Company was a country store established by William M. Barnwell on June 14, 1926 on a piece of land known as the "Hicks Tract" of the Roxbury Plantation. Darby Jones managed and lived behind the store from some time prior to 1928 until 1945, when Barnwell's son Ozzie took over operation of the business. He and his family lived behind the store until 1955. Ozzie eventually became the owner as well, until his death in 1979, when the store was operated by Hugh Allston until in burned in 1983. Below is a picture of Ozzie having some fun while riding a tricycle around inside the store. (Both, courtesy of Joyce "Dumpy" Towles Barnwell Cave.)

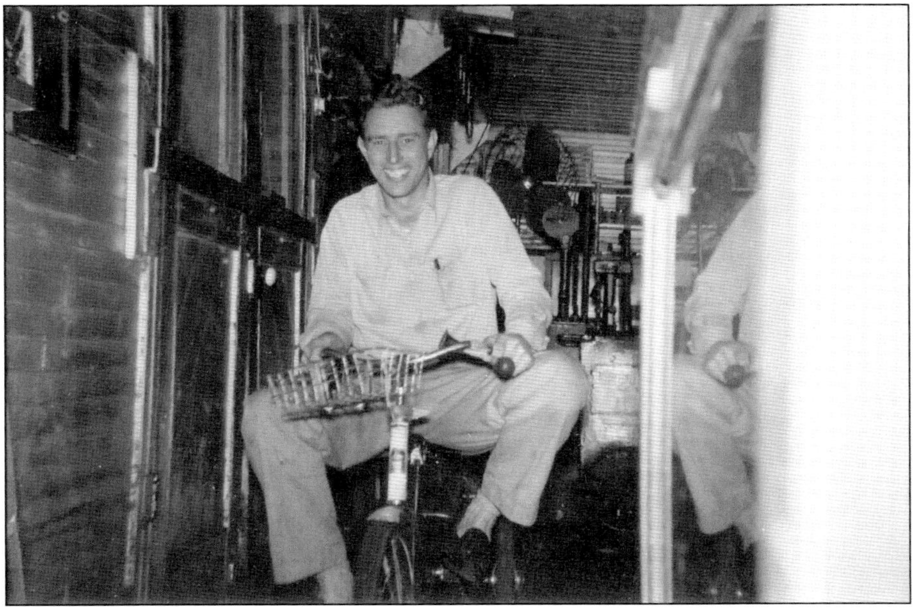

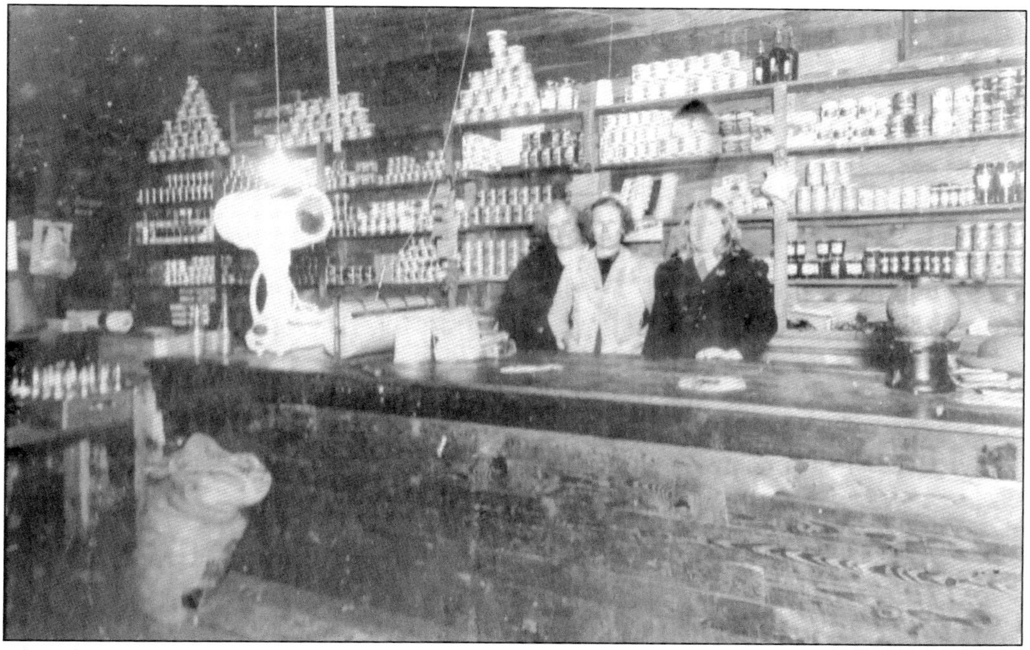

This photograph taken in the mid-1920s shows three ladies behind the counter at Hallex's Store in Ravenel. The store, as many others in the area, sold dry goods, food, and sundries to the residents who lived nearby. (Courtesy of Buddy Hill.)

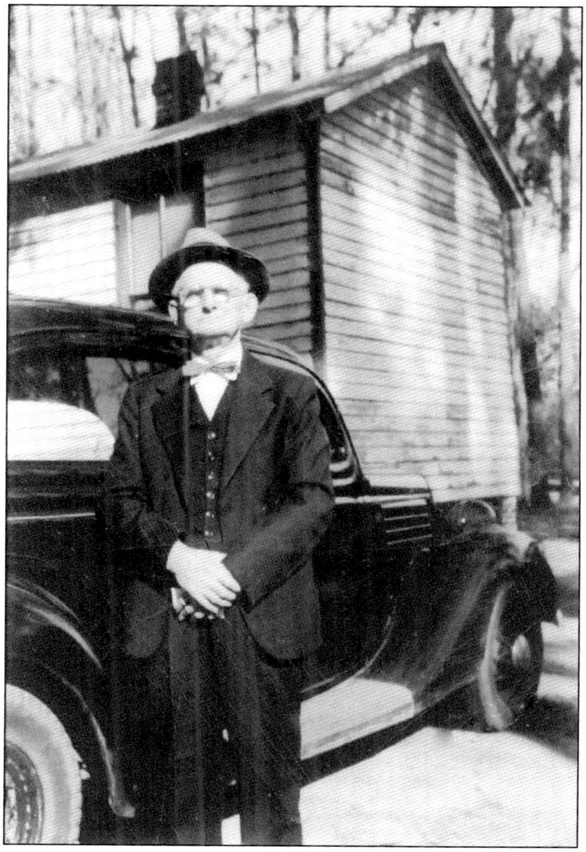

Arthur Kearse, a merchant who ran a store called "The Juck" in the heart of the village of Adams Run, is pictured around 1924. The Juck was one of five stores in the village and primarily sold food to local residents. Kearse was originally from Ehrhardt, South Carolina, and he was the cousin of local well-to-do farmer Gordon Kearse. (Courtesy of Jackie Boineau.)

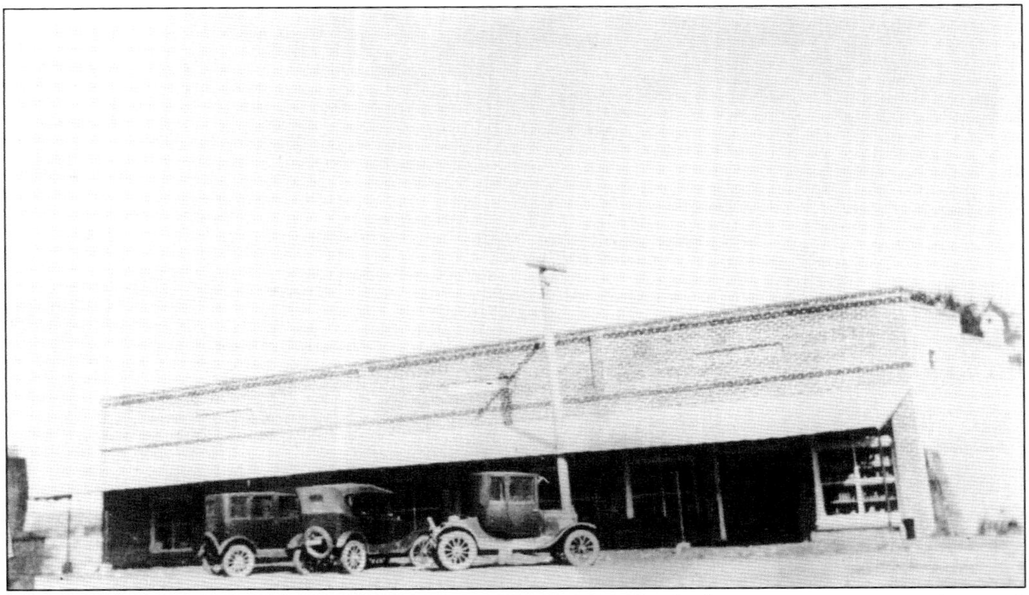

As in many small towns, the shops along Main Street served a dual purpose: a meeting place and a place to buy needed goods and services. These storefronts in downtown Meggett are shown around 1920. (Courtesy of Buddy Hill.)

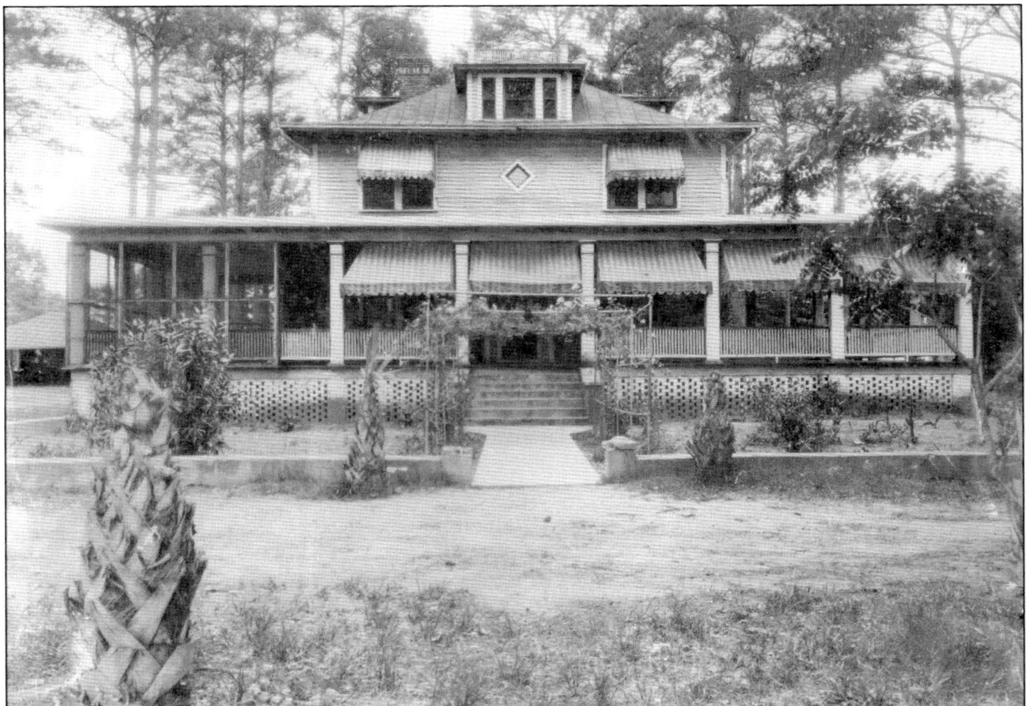

The Quigley House was the Towles family home. The house with its wide porches provided a gathering place for family and friends. The home remains in the family today and is virtually unchanged. (Courtesy of Beth Towles Van Norte.)

In the early 20th century, the Argyle (formerly known as the St. Charles), one of Charleston's three leading hotels, was located at the corner of Meeting and Hasell Streets. The hotel earned a reputation for itself during the West Indies Exposition of 1901–1902 for its elegance and modern amenities and is believed to have been used as a backdrop for Donald McCaig's 2007 novel *Rhett Butler's People*. While the hotel's 18 Ionic wrought-iron columns were an architectural symbol of Charleston's antebellum past, the interior spaces housed many an important meeting. The Argyle was used often by the truck farmers of Meggett as a place to get together, talk business, play poker, and socialize. At one point, the hotel was purchased by the truck farmers of St. Paul's Parish, and millionaire farmer Frank E. Towles acted as director of the hotel for some time. (Right, courtesy of Beth Towles Van Norte; below, courtesy of Jennifer Gilliland.)

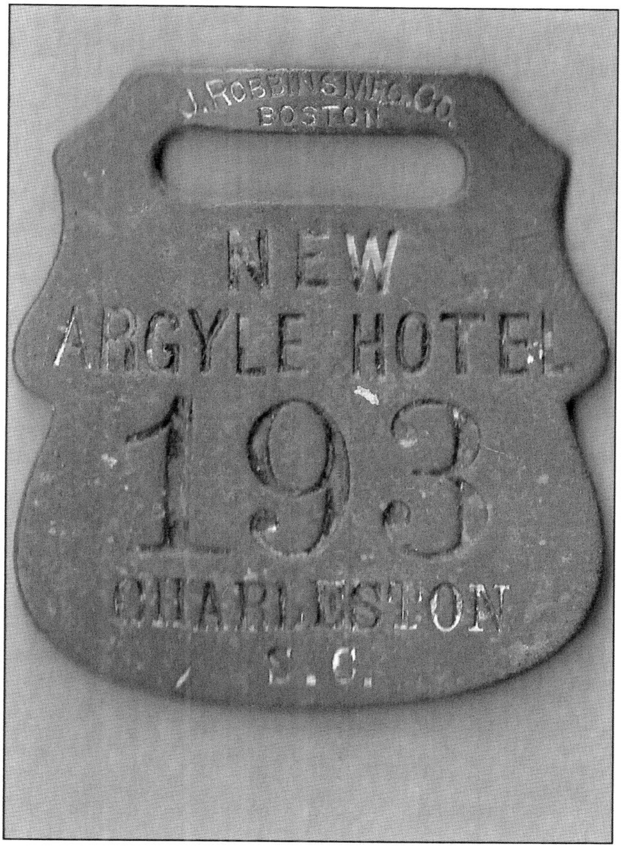

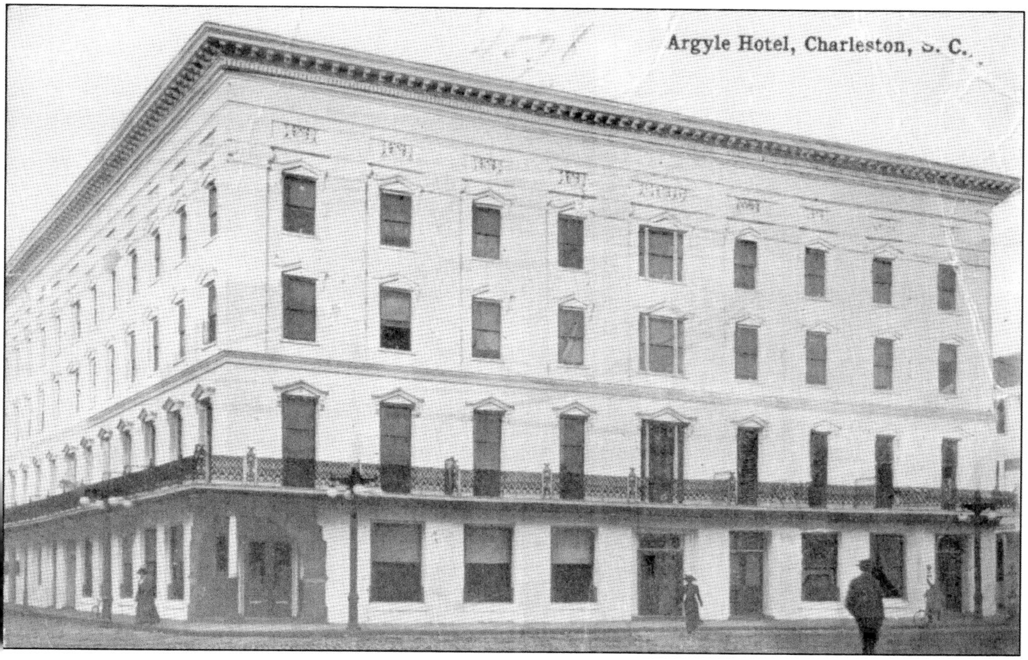

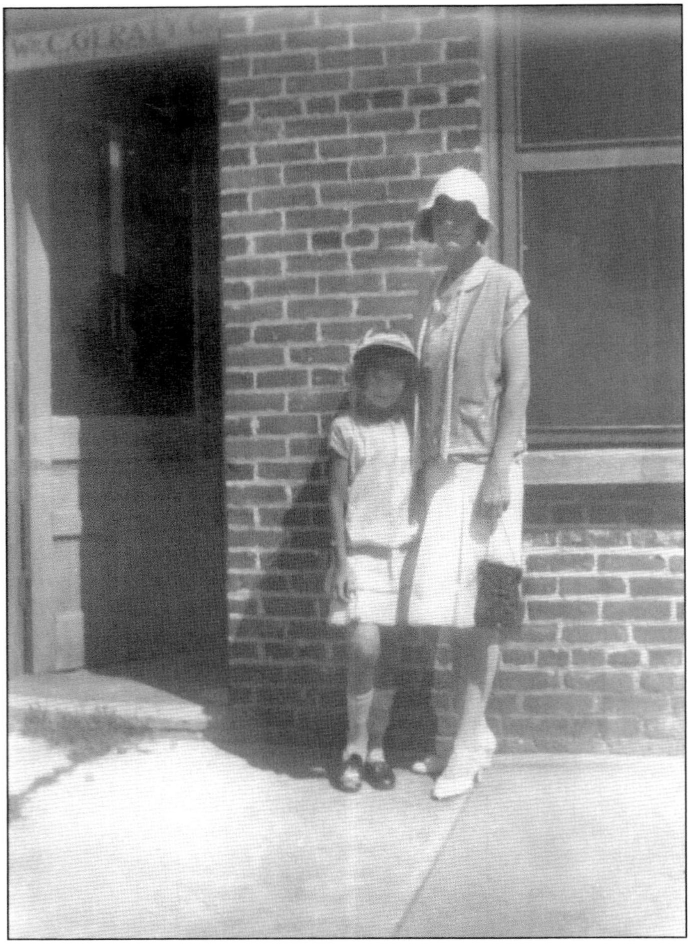

Antoinette "Net" Griner (left) and her aunt Helen Griner Mays pose in front of the two-story brick building that at one time was the Yonges Island Post Office and headquarters for W.C. Geraty and Company. The photograph is believed to have been taken around 1920. (Courtesy of Anne Sanders Groover.)

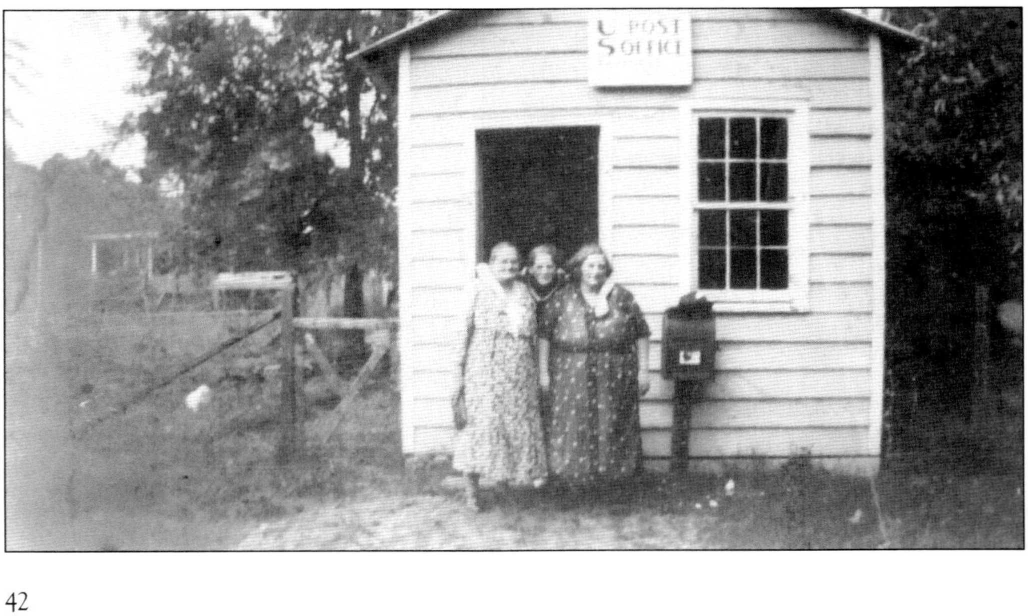

Three unidentified ladies stand in front of the original frame Ravenel Post Office sometime in the early 1920s. Several years later, the Ravenel Post Office became the official courthouse in the town. (Courtesy of Buddy Hill.)

Pictured are postmistress Florie Linder Smith and a young Jackie Boineau (age six) at the Osborn Post Office in 1936. Florie was married to Jack Smith, and her brother was a supervisor at the nearby barrel factory. (Courtesy of Jackie Boineau.)

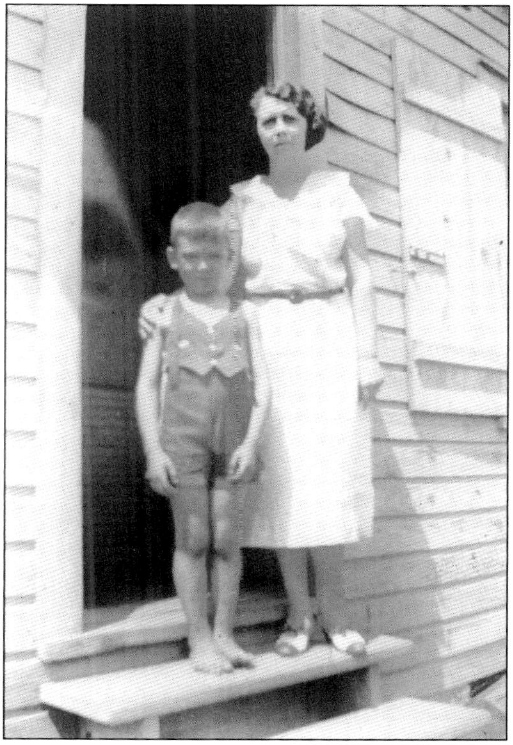

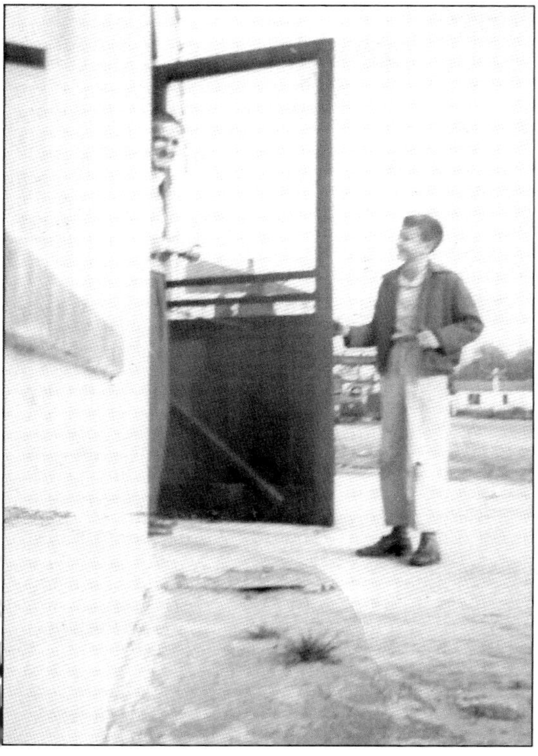

A young Ronald "Ronnie" Rash (age 14) stands with postmaster Laurie Sandifer in front of the Hollywood Post Office in 1955. The post office seen in this picture is part of the current-day Hollywood Town Hall. (Courtesy of Ron Rash.)

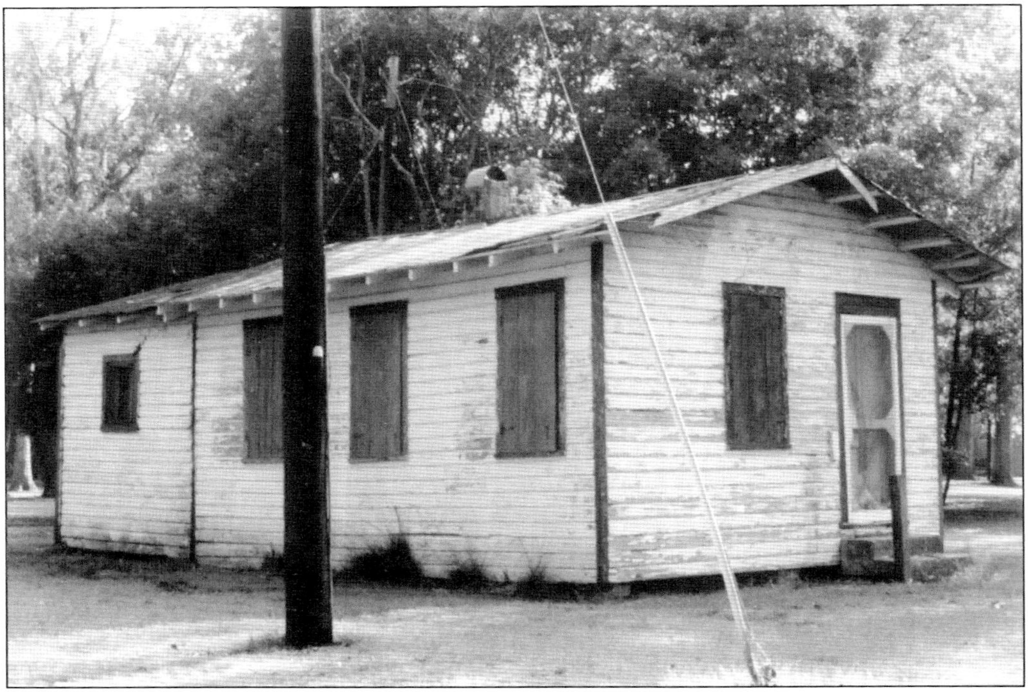

Above is the original Adams Run Post Office, which dates to the early 1930s. It was located on a dirt road near the grammar school in the Adams Run village. At the time, there was no home delivery, so everyone went to the post office daily to collect mail. Gordon Taylor (among others) was the postmaster at this office. Seen below is the current Adams Run Post Office located on Highway 174 in Adams Run. (Both, courtesy of Jackie Boineau.)

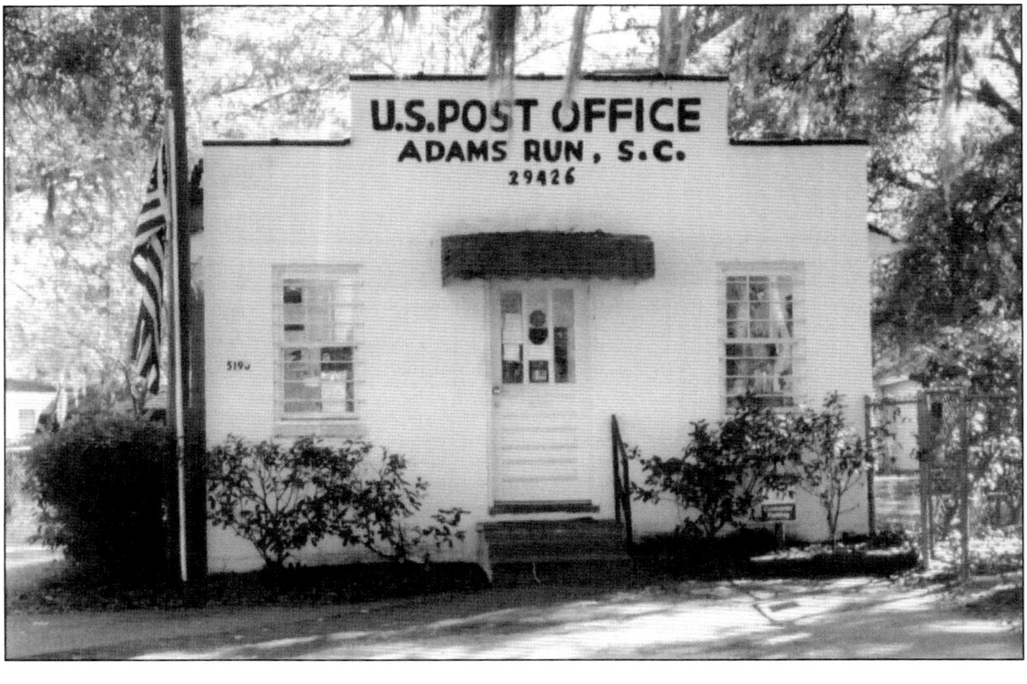

Seen at right is Carl Jefferson Butler Jr. (September 19, 1930–October 1, 2009), otherwise known to the residents of Hollywood as "Mr. B," "Carl Jr.," or "Mr. Carl." He is standing against a gas tank outside of his Butler's Gulf Station in 1957. Butler's started off as a grocery store when it was owned by Carl's father, Carl Jefferson Butler Sr., but later became a gas station that offered more than just gas to the residents of Hollywood. Sandwiches and desserts were sold there, and the business became a gathering place for old and young alike before and after work. Below is a picture of Hollywood looking down the street from Butler's. (Both, courtesy of Mona Butler Williams.)

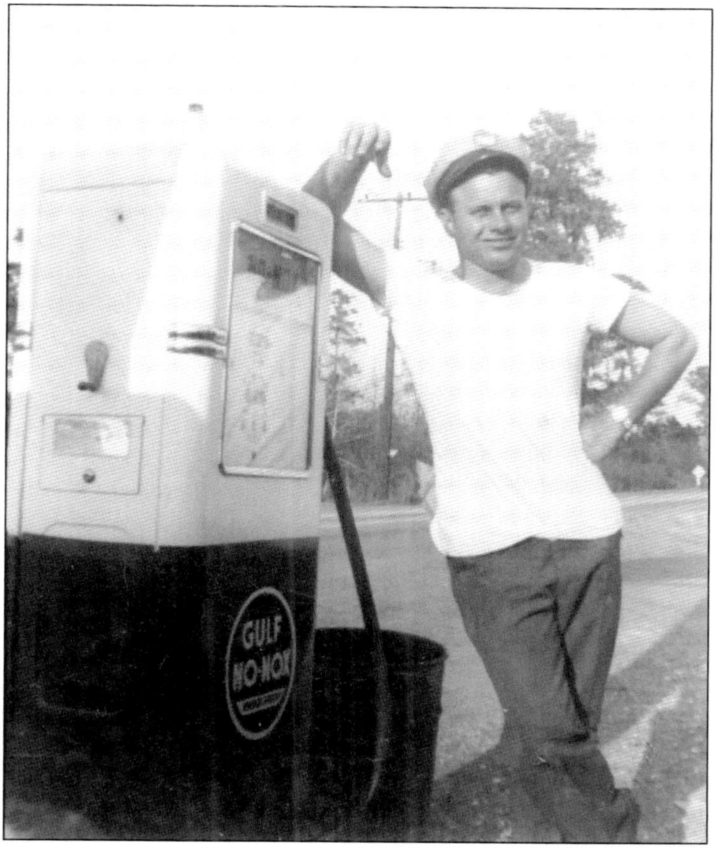

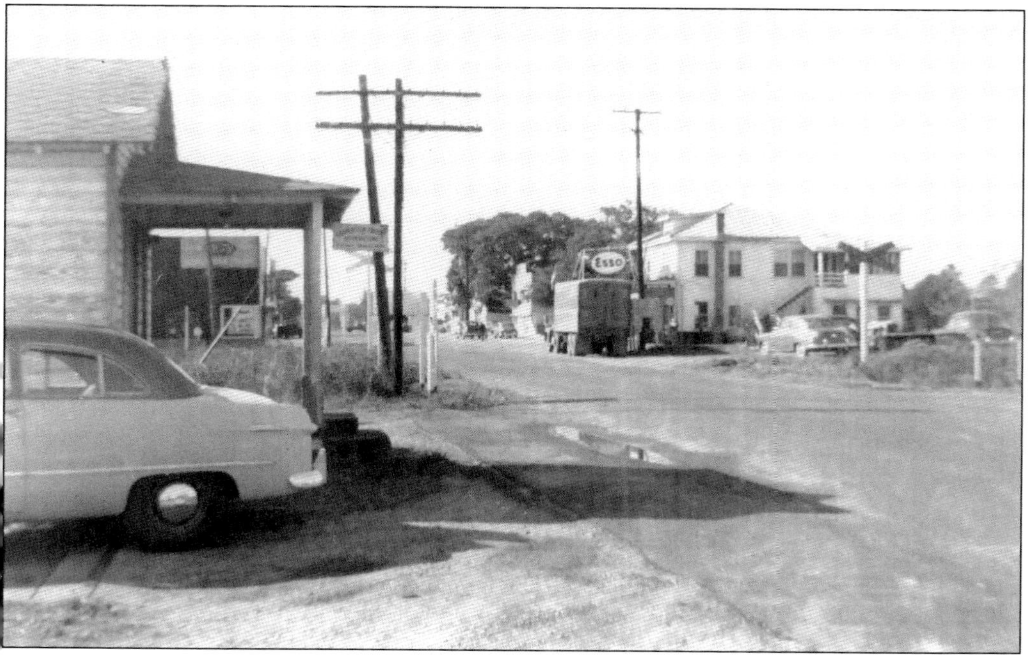

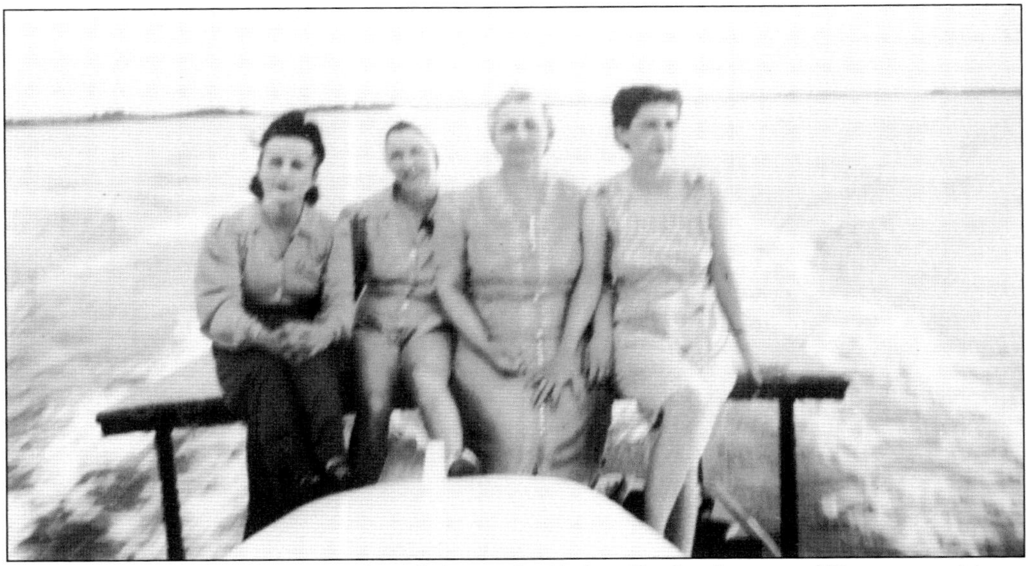

Seen here are the Butler sisters—from left to right, Esther, Emilee, Lena, and Frances—taking a ride on their father's (Ferdinand Winfield Butler) boat on the Toogoodoo River in St. Paul's Parish. Boating was and continues to be a favorite pastime among residents in the parish. (Courtesy of Mona Butler Williams.)

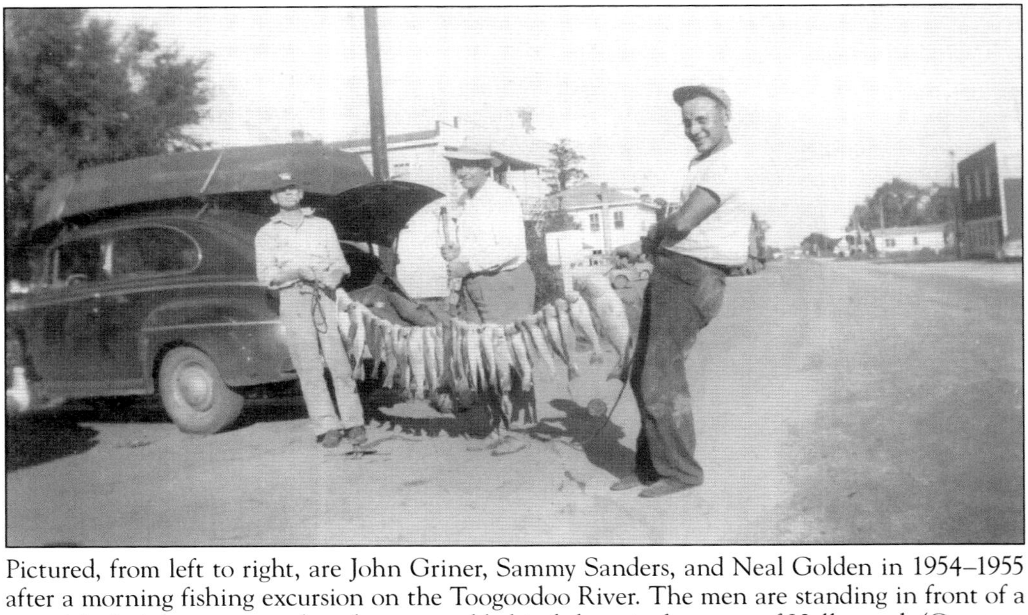

Pictured, from left to right, are John Griner, Sammy Sanders, and Neal Golden in 1954–1955 after a morning fishing excursion on the Toogoodoo River. The men are standing in front of a car with their boat strapped to the top, and behind them is the town of Hollywood. (Courtesy of Anne Commins.)

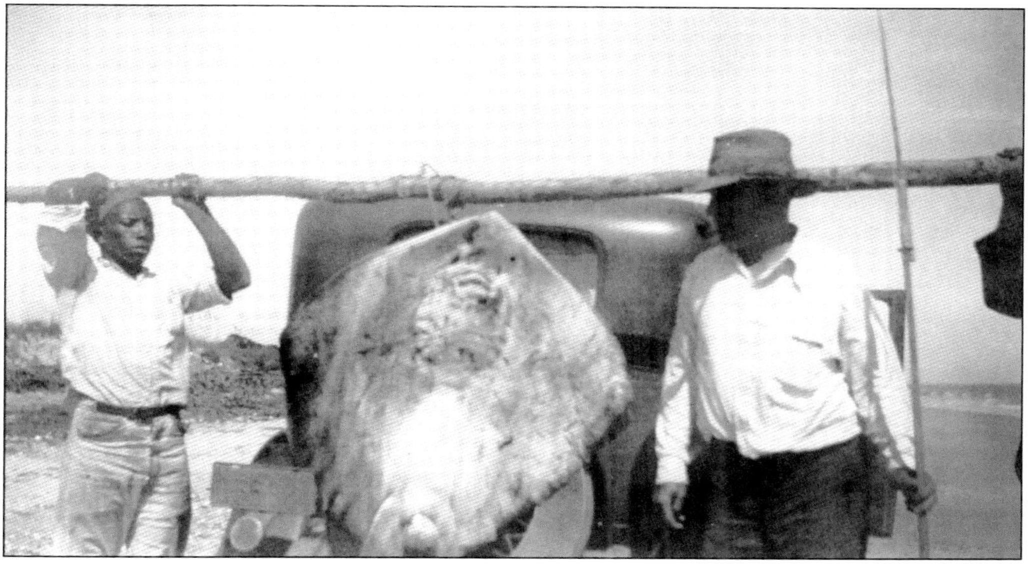

A very large stingray that had been captured or found caught in the tide along one of the larger rivers in the parish is seen in this 1920s photograph. In the photograph, an unidentified young African American male is helping Dr. Prentiss (right) secure the catch. Stingrays inhabit the ocean waters off of South Carolina but are also believed to make their way up some of the larger rivers that run through the barrier islands. (Courtesy of Marsha Blandenburg.)

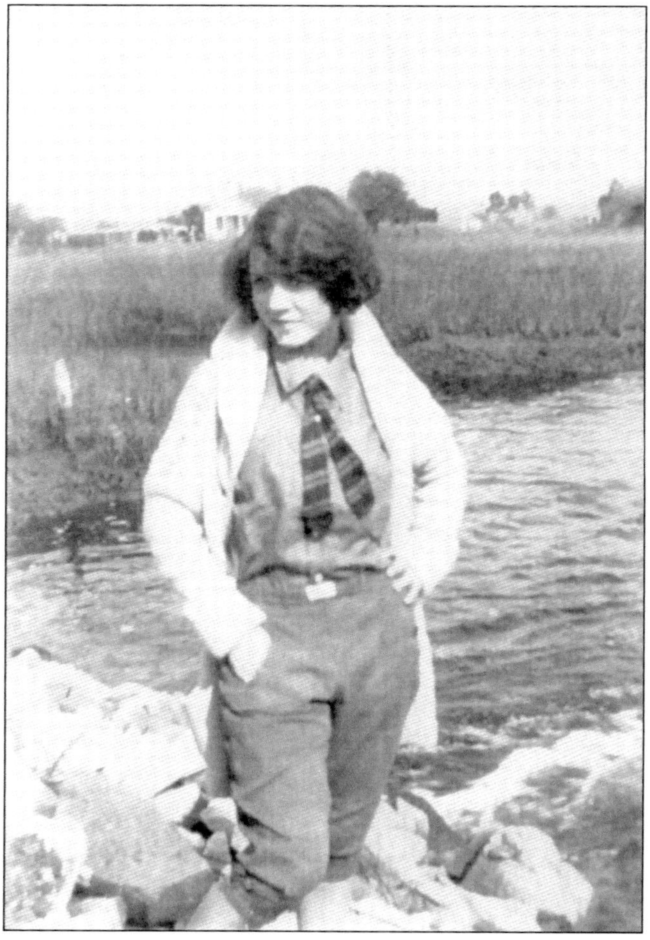

In 1925, a young Natalie Searson Towles stands on some rocks along the river showing off a new outfit purchased by her father. Natalie was just 15 and a student at the College of Charleston. One of the youngest students ever to attend, she was just 14 years old when she began college. (Courtesy of Natalie Towles Whiteman and Liz Whiteman.)

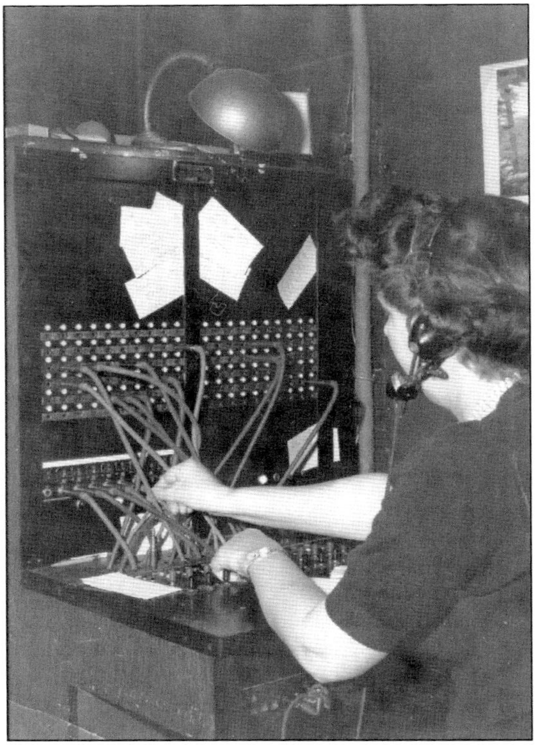

Seen at left is Zelma Blackburn Rash in the 1950s working the Hollywood Telephone Exchange, known simply as the switchboard. Rash worked 16-hour days connecting local callers, placing long-distance calls, and even sounding the alarm for the fire department. She slept on a roll-away bed next to the switchboard in the event that there was an emergency, and her children shared the one bedroom in the back of the office. Seen below is Myrtie Smith, of Adams Run, who was hired to work an eight-hour shift to help Rash. (Both, courtesy of Ron Rash.)

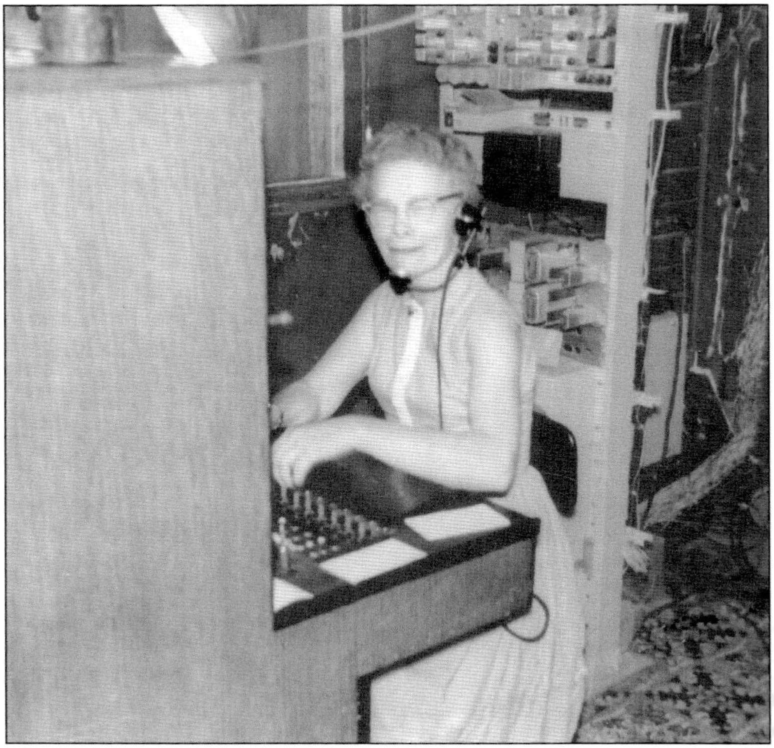

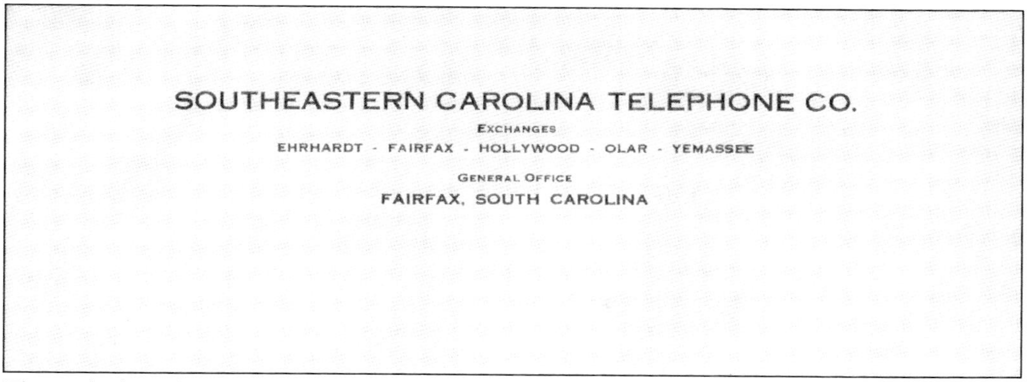

SOUTHEASTERN CAROLINA TELEPHONE CO.

EXCHANGES

EHRHARDT - FAIRFAX - HOLLYWOOD - OLAR - YEMASSEE

GENERAL OFFICE

FAIRFAX, SOUTH CAROLINA

This is the letterhead of the Southeastern Carolina Telephone Company that was the predecessor to what is now Frontier Communications Company. The local exchanges were Ehrhardt, Fairfax, Hollywood, Olar, and Yemassee. The exchange in Hollywood served as the main hub for all telephone communications for the parish. (Courtesy of Ron Rash.)

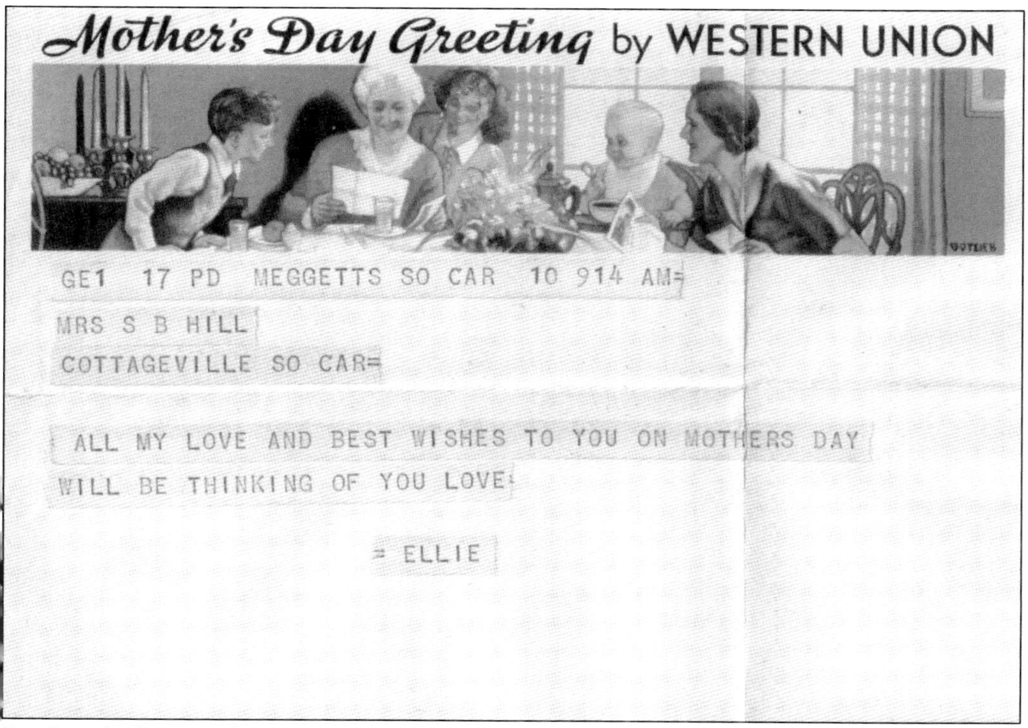

Mother's Day Greeting by WESTERN UNION

GE1 17 PD MEGGETTS SO CAR 10 914 AM=

MRS S B HILL

COTTAGEVILLE SO CAR=

ALL MY LOVE AND BEST WISHES TO YOU ON MOTHERS DAY

WILL BE THINKING OF YOU LOVE

= ELLIE

Besides telephones, telegrams remained a popular form of communication in towns, and Meggett was no exception. This telegram was sent from a resident of St. Paul's Parish to the sender's mother in Cottageville on Mother's Day; it is dated May 10, 1945. (Courtesy of Linda Allston.)

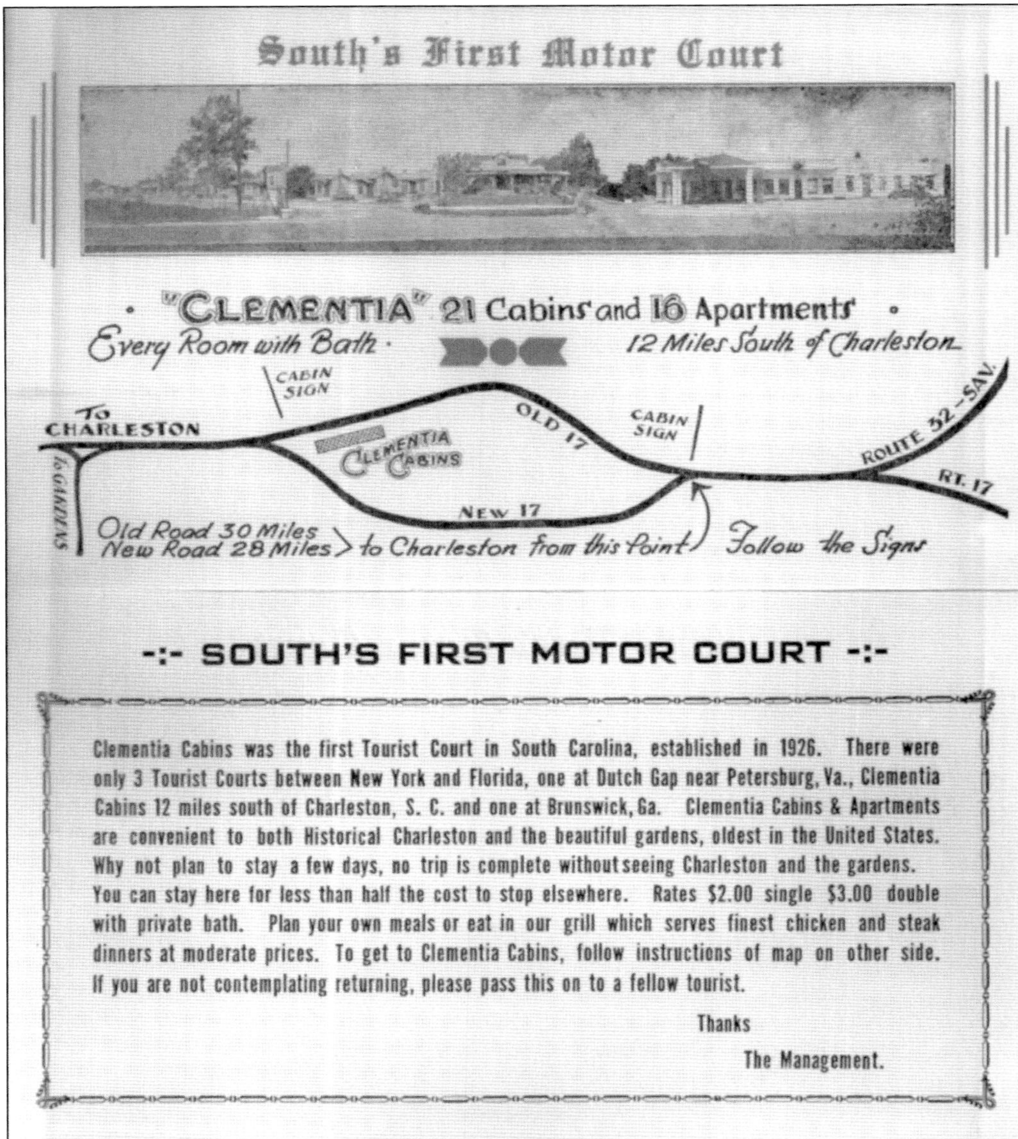

This is an advertisement for Clementia, which was one of only three motor courts between Florida and New York in the 1920s. Clementia was established in 1926 and featured "cabins" with private baths surrounding the main restaurant and nightclub. Clementia was located on what is now Highway 162 in St. Paul's Parish, which at the time was the main highway leading into and out of Charleston. When the new US Highway 17 was constructed, it was only a short time before Clementia started to struggle because of the loss of car traffic. The old motor court burned and is no longer standing. (Courtesy of Wilbur Jones.)

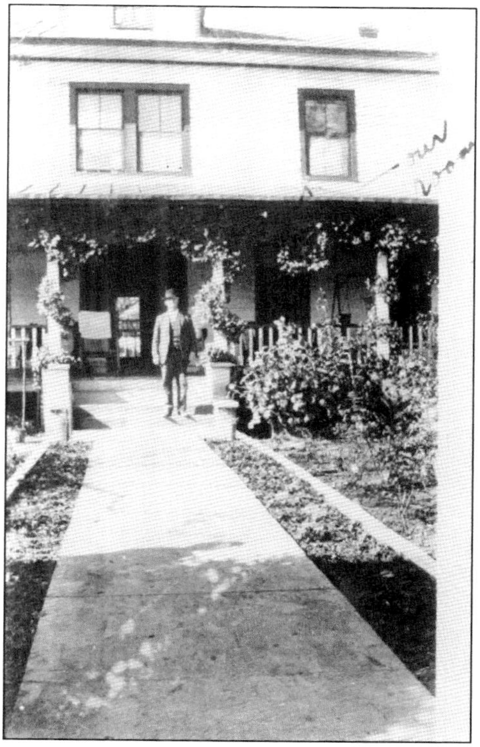

At right is the Gay family boardinghouse that was located adjacent to the Meggett Depot. The inn was home to many of the single teachers at the Meggett School and other young professionals who worked in and around the Meggett area. In the photograph below, taken around 1926, are several of the residents at a wedding reception held at the boardinghouse. From left to right are Robert Platt, Miss Rogers, young Jimmy Gay, Miss Derrick, Olive Cox, and Miss Johnson. Rogers, Johnson, Derrick, and Cox were all teachers at the school. (Both, courtesy of John Waring.)

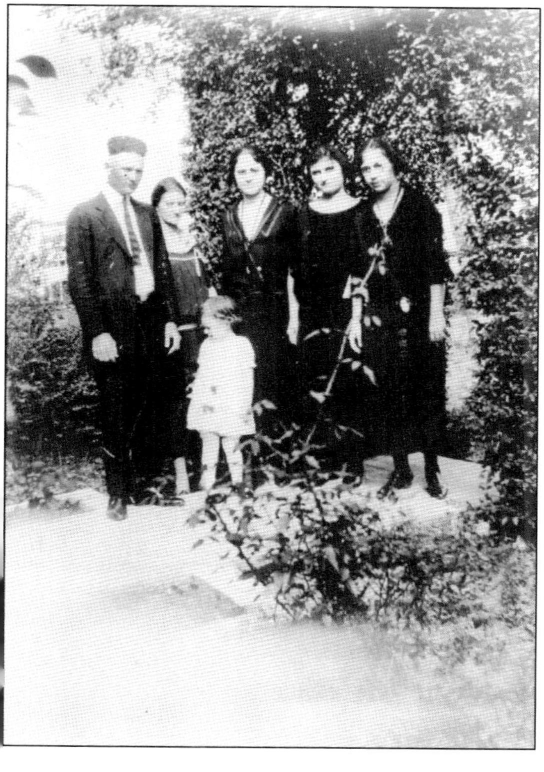

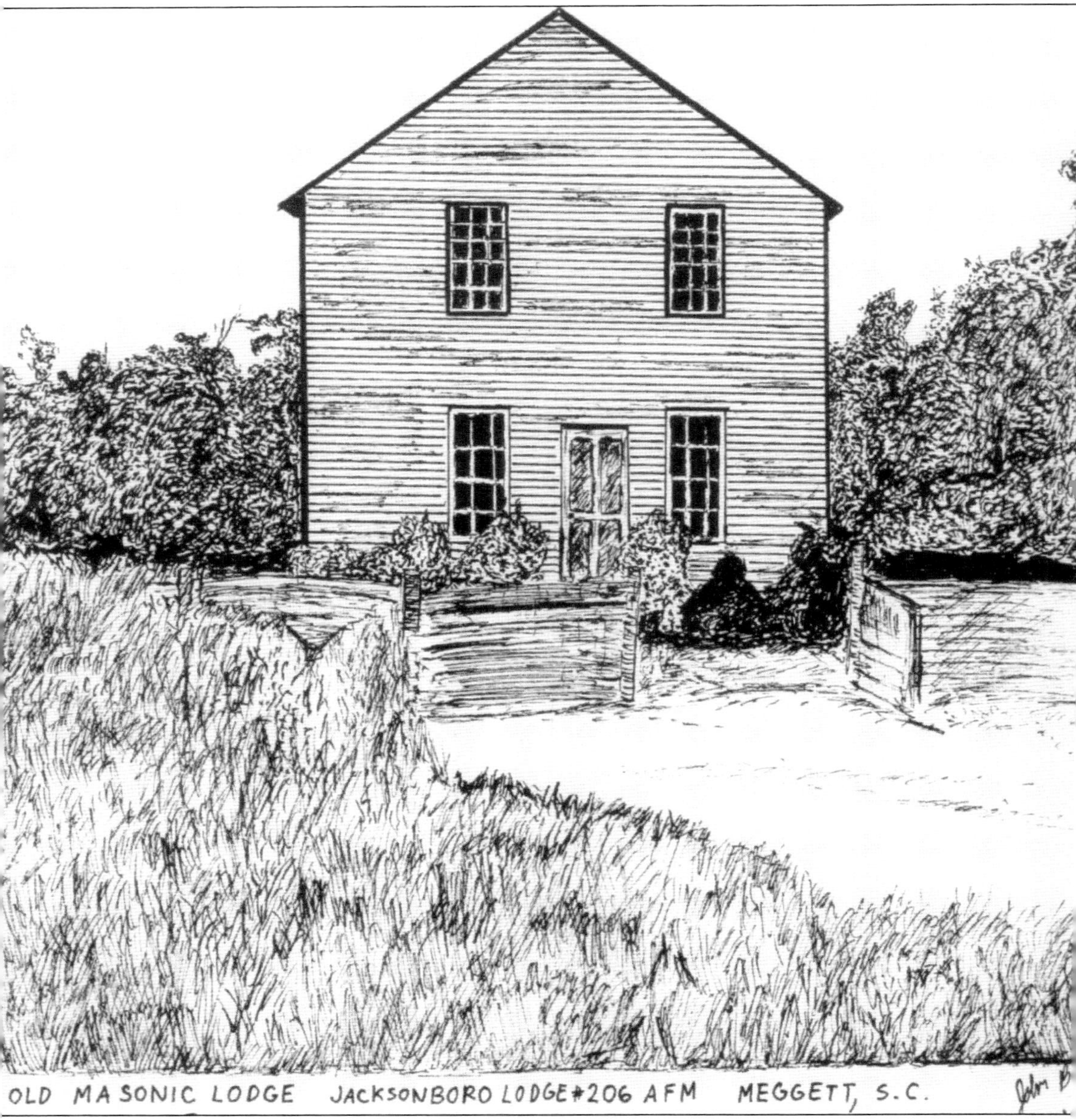

Seen in this rendering by artist John Waring is the old Masonic Lodge that is fondly called the "Chicken House" by many Meggett residents. The structure is believed by many, including Beulah Glover, to be the Jacksonboro Masonic Lodge that was once the meetinghouse of the South Carolina House of Representatives when Jacksonboro was the provisional state capital for 40 days from January 8 to February 26, 1782. (Courtesy of John Waring.)

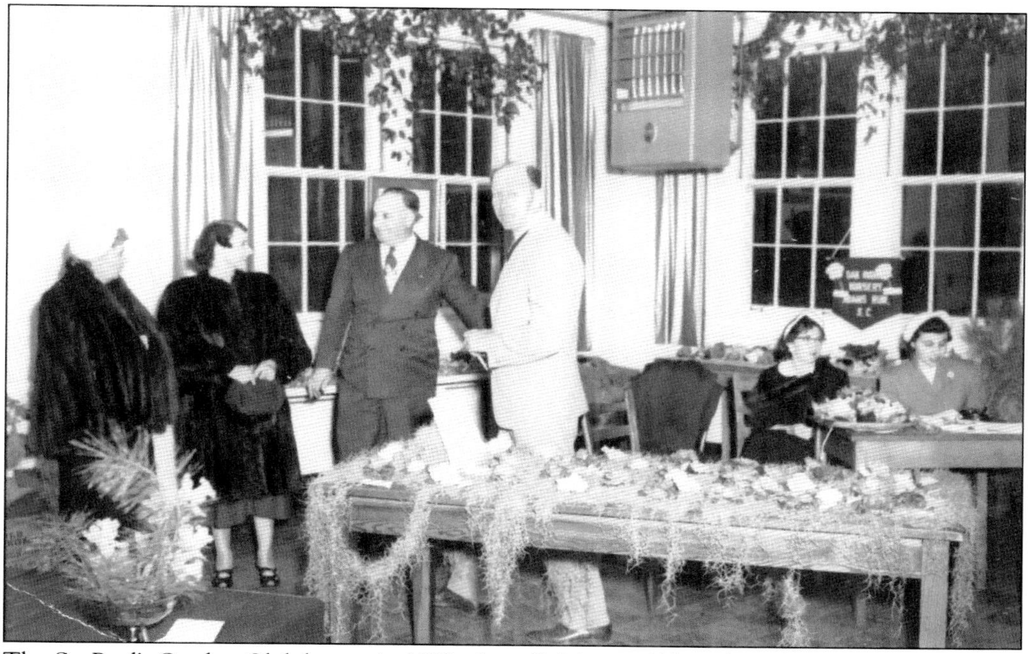

The St. Paul's Garden Club began in 1929 when O. Morton Edwards and his wife, Bertie (both of whom were teachers at St. Paul's High School) decided to get the community involved in gardening and agriculture. The club was official and federated in 1932 and continues to be active today. Over the years, the club and its members have been leaders in improving the community—from victory gardens during World War II to the beautification of local school yards. In the photograph are, in no particular order, Claude Prentiss and Mr. and Mrs. Herman Dryer receiving an award at the Garden Club Flower Show held at the Hollywood Community Center. (Courtesy of Marsha Blandenburg.)

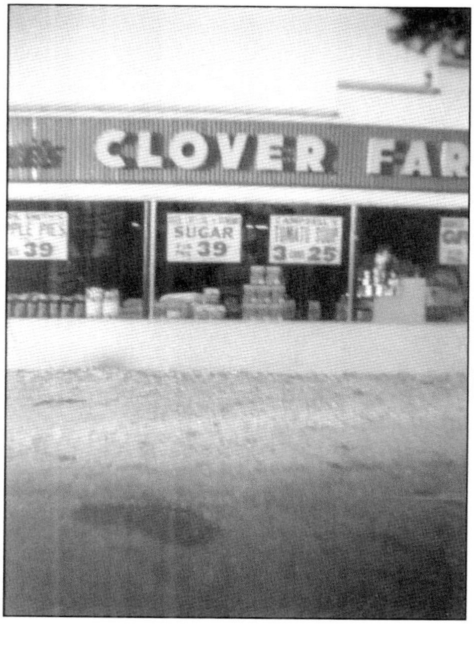

Clover Farms, Hollywood's first supermarket, is pictured around 1955. It was owned and operated by John Hardee and was in the center of town where Arrow Cleaners is located today. (Courtesy of Ron Rash.)

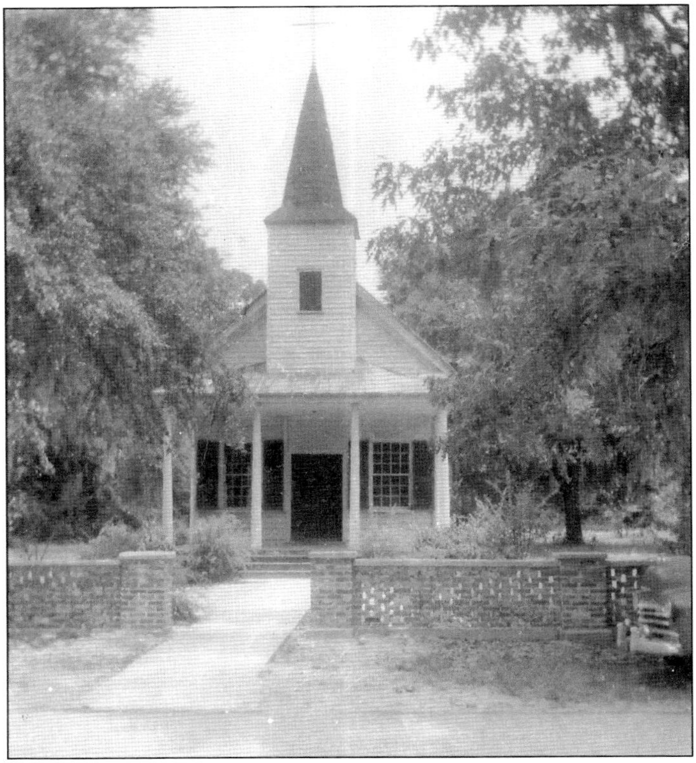

At left is a picture taken outside of Christ Church in Adams Run in 1972. Below is the interior of the church on Easter Sunday in the mid-1950s. The Episcopal church in St. Paul's Parish dates back to the Church Act of 1706 when the Church of England was established in the colony of South Carolina. The Episcopal church remains a vital part of the community today. (Both, courtesy of Linda Allston and Rosanne King Jones.)

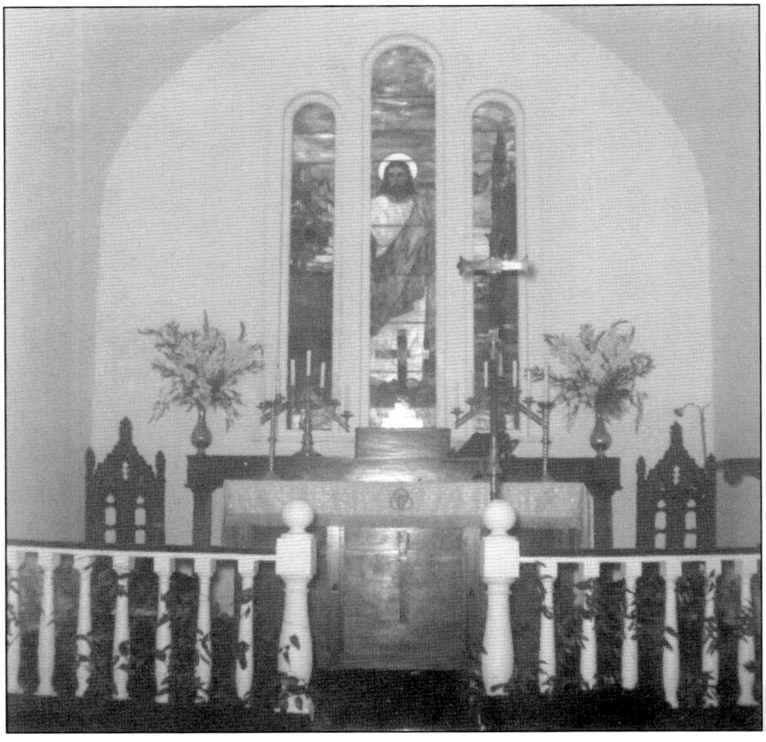

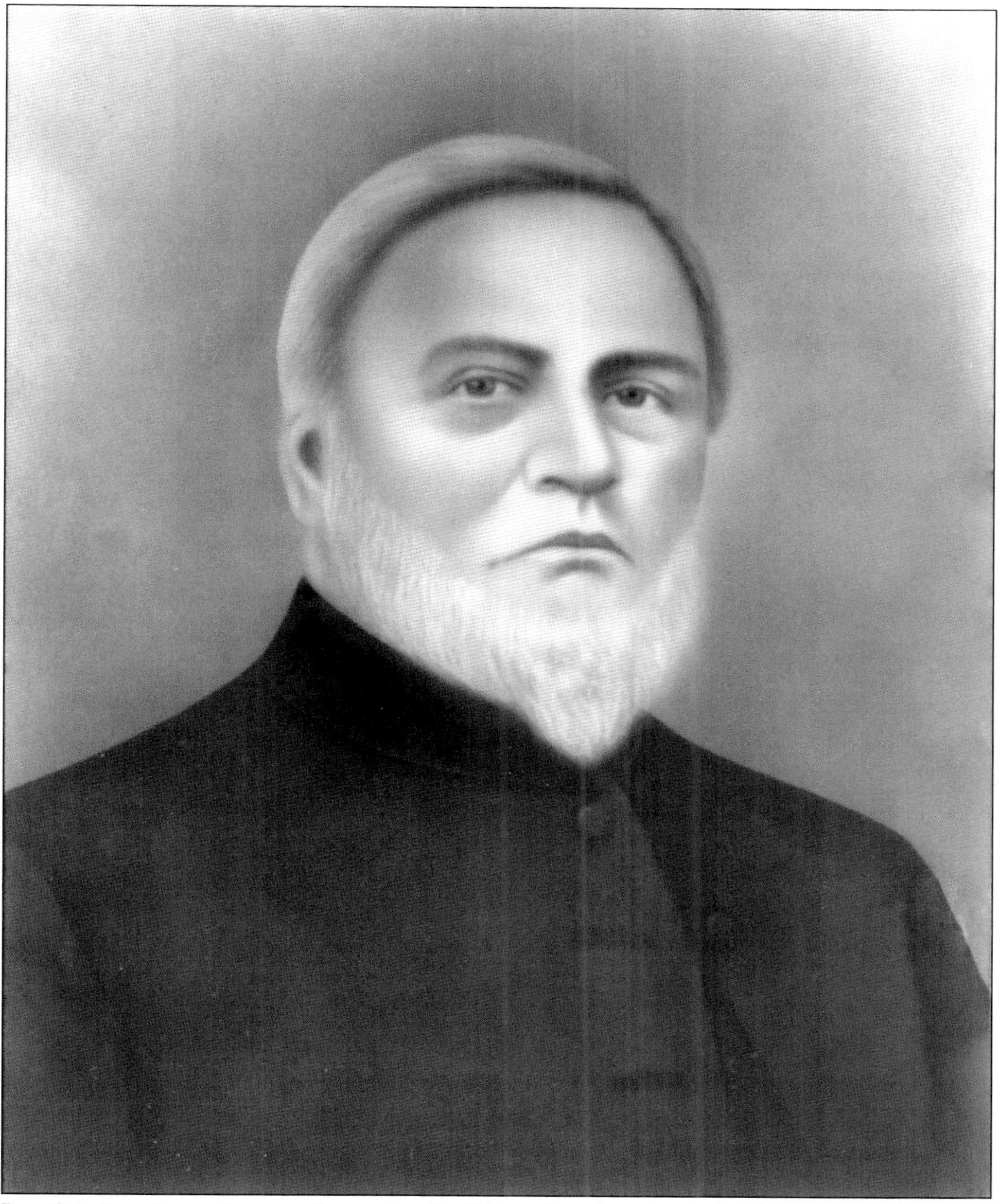

Rev. Otis Prentiss was born on September 30, 1814, in Beaufort, South Carolina. Reverend Prentiss was an Episcopal clergyman who served many years in around St. Paul's Parish. He graduated from the College of Charleston in 1833. Reverend Prentiss served at Christ Church in Adams Run from 1853 to 1857 and from 1868 to 1893. (Courtesy of Linda Allston.)

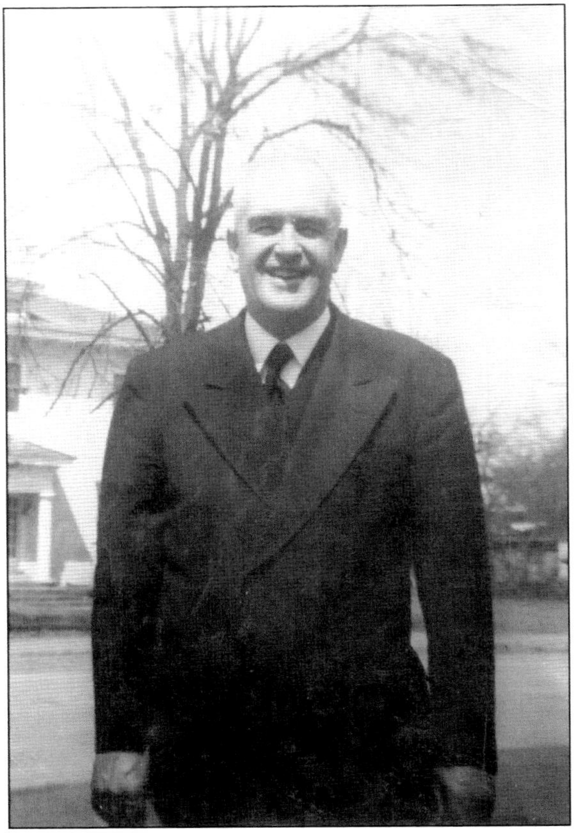

In this picture from around 1940 is Rev. Sumner Guerry. He was the rector of Christ Church in Adams Run from April 1935 to January 1944. In addition to serving as the reverend for Christ Church in Adams Run, he also was a rector of an Episcopal church in North Charleston. Guerry never married and is buried in the St. Phillip's Church graveyard in downtown Charleston. (Courtesy of Edward Guerry and Jackie Boineau.)

These are the children attending Vacation Bible School in 1952 at Christ St. Paul's in Yonges Island. (Courtesy of Rosanne King Jones.)

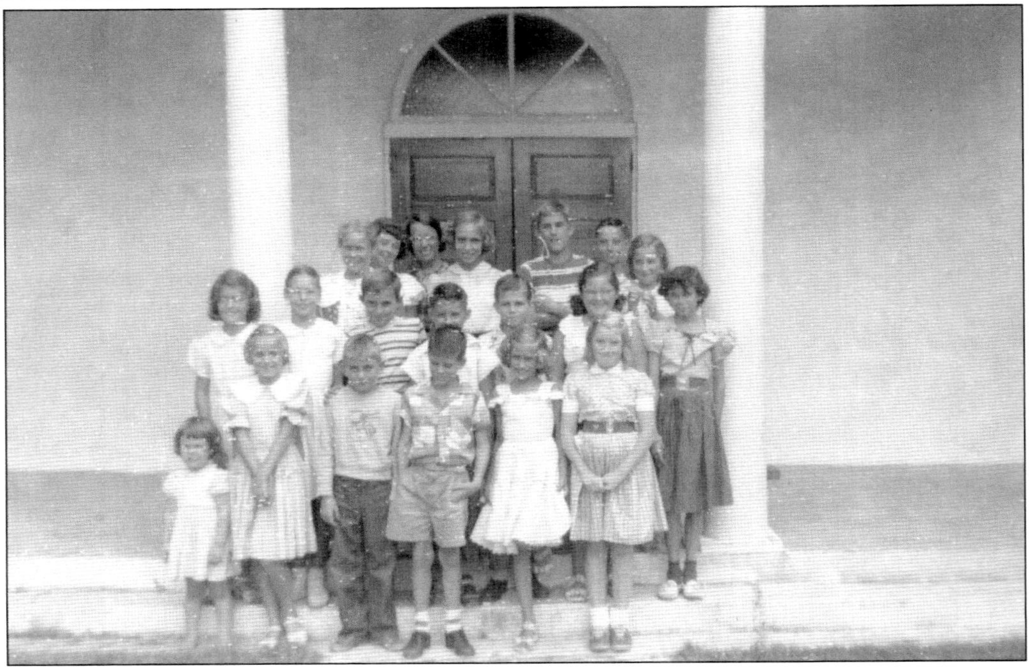

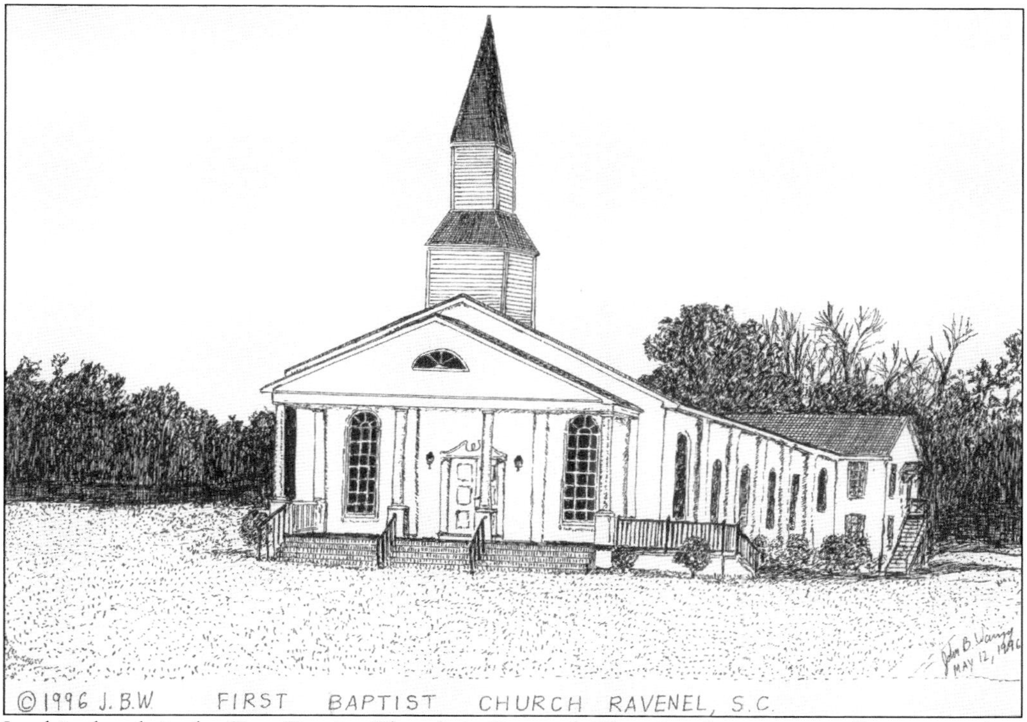

In this sketch is the First Baptist Church in Ravenel. The church was originally organized in 1913 and situated near the railroad tracks on Martin Street in Ravenel. In 1957, the congregation moved to its present location at the intersection of Highways 17 and 165. Many materials used to build the present-day church were obtained from historic buildings and churches in Savannah and Charleston, including the marble flooring in the pulpit, vestibule, and on the portico, which came from the historic Argyle Hotel in Charleston. (Courtesy of John Waring.)

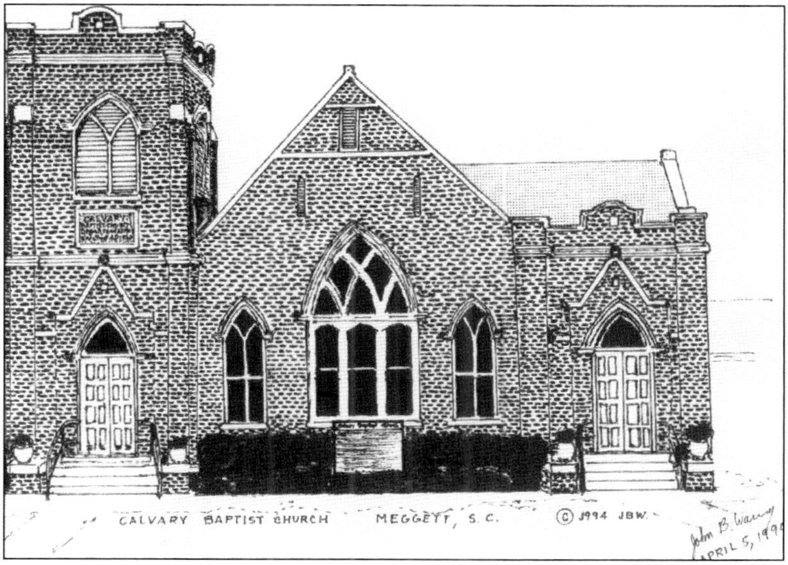

This sketch shows Calvary Baptist Church, located in Meggett. The organization of the church dates to 1904 when it was established by a group of local residents. Today, Calvary Baptist continues as a thriving congregation in the community. (Courtesy of John Waring.)

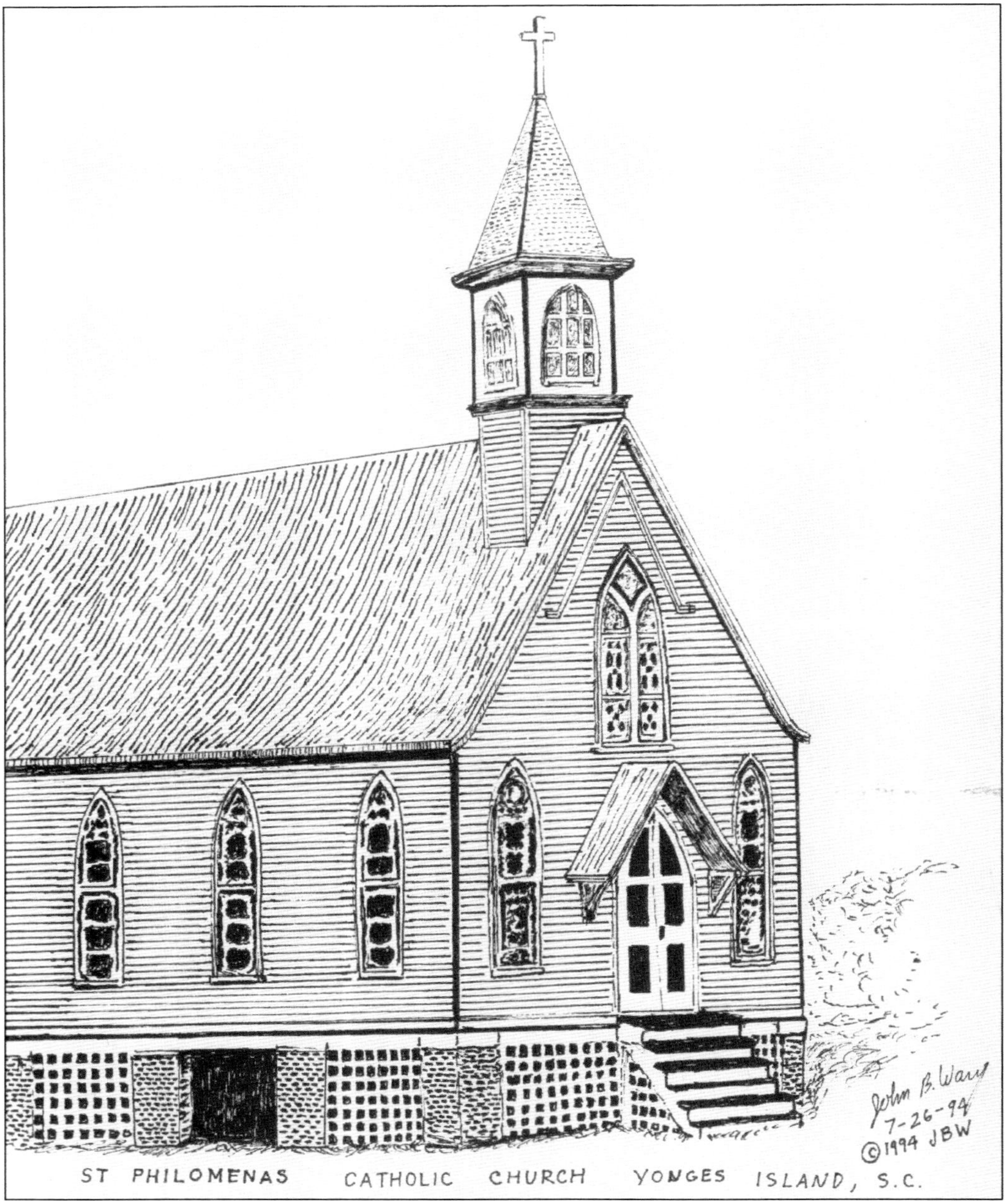

Shown in the sketch is St. Philomena's Church, which was a Catholic congregation started by friar Daniel J. Quigley at Martin's Point on Wadmalaw Island. The first marriage at St. Philomena's was recorded on February 9, 1872, between William Christopher Geraty and Sarah Anne Ray. Because of the increasing population on Yonges Island, it was decided to move the church to that area, and on May 12, 1885, the church was dedicated in its new location. (Courtesy of John Waring.)

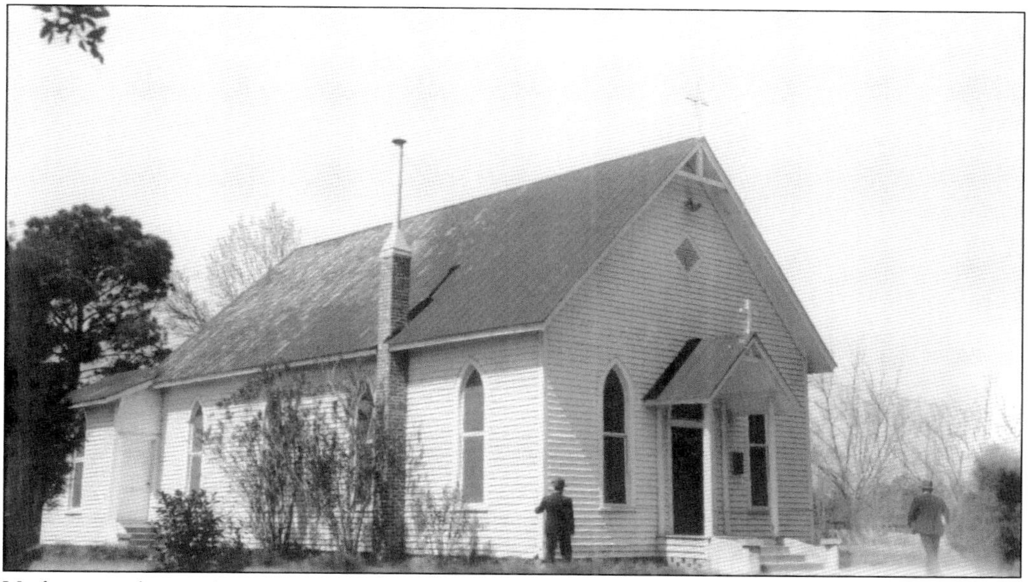

Unfortunately, on August 27, 1911, a hurricane destroyed the original St. Philomena's Church, except for the stained-glass window. It took three years to rebuild, but the new church was finally completed in April 1914. There had been a few changes, one of which had been the renaming of the church as St. Mary's. (Courtesy of St. Mary's Church.)

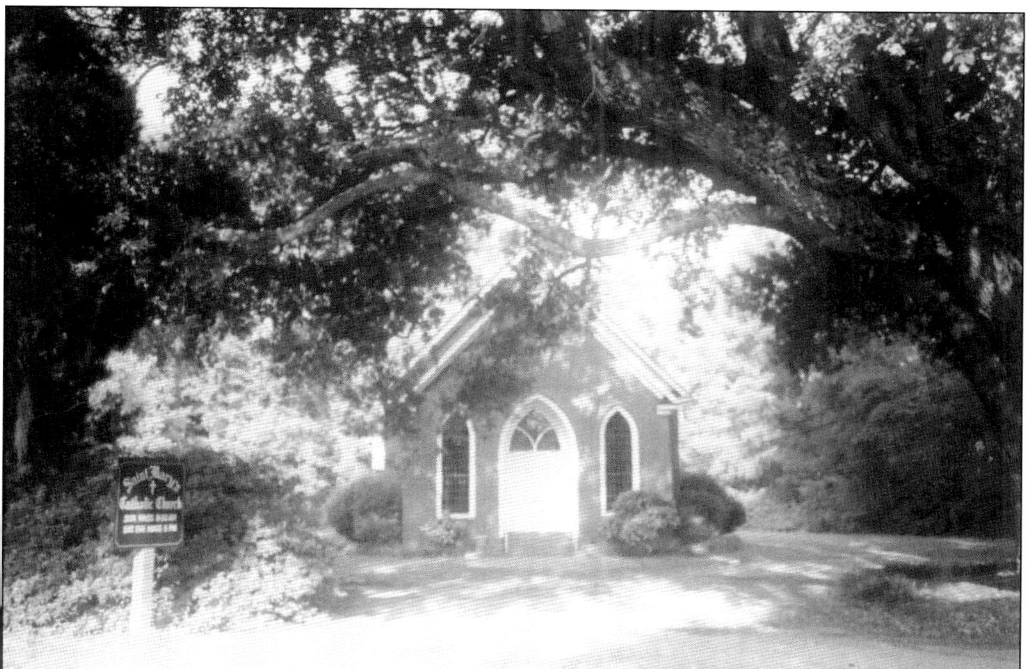

St. Mary's Catholic Church is on Yonges Island. The exact date of when the clapboard church was bricked over is not known; however, the structure as seen here is still standing and has an active congregation today. (Courtesy of John Waring.)

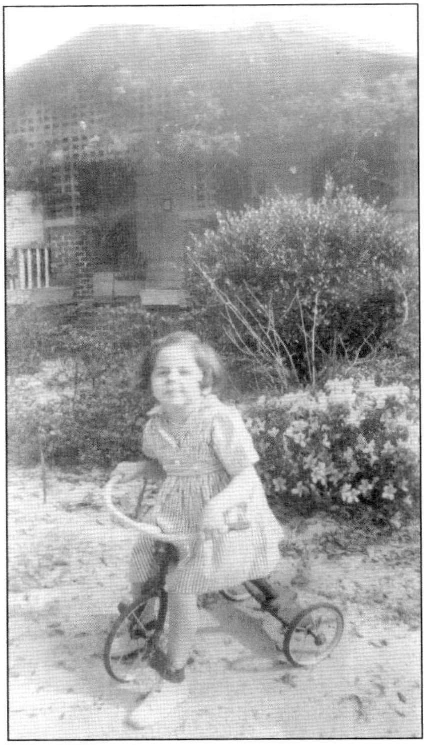

Marie Cynthia Sanders (1939–1993) was the daughter of Net Griner and John Samuel "Sammie" Sanders. Marie, who suffered from cerebral palsy, loved to write short stories, some of which were published during her lifetime. Marie is perhaps best known in the community for her written history of St. Mary's Church on Yonges Island. (Courtesy of Anne Sanders Groover.)

Below is a picture of St. Mary's Catholic Church's confirmation class taken in 1954–1955. From left to right are (first row) Johnny Knisley, Anne Sanders, and Jim Moody; (second row) Steve Flowers, Jimmy Geraty, and Jackie Butler. Father Sheady is standing behind the children. (Courtesy of St. Mary's Church.)

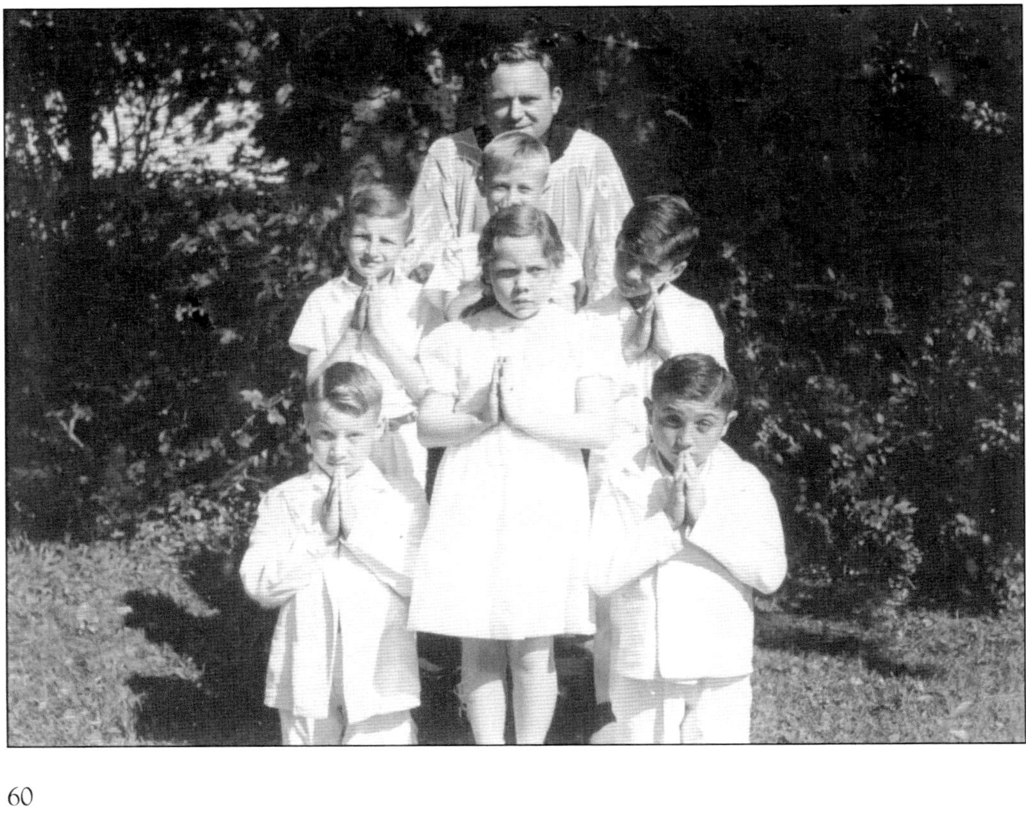

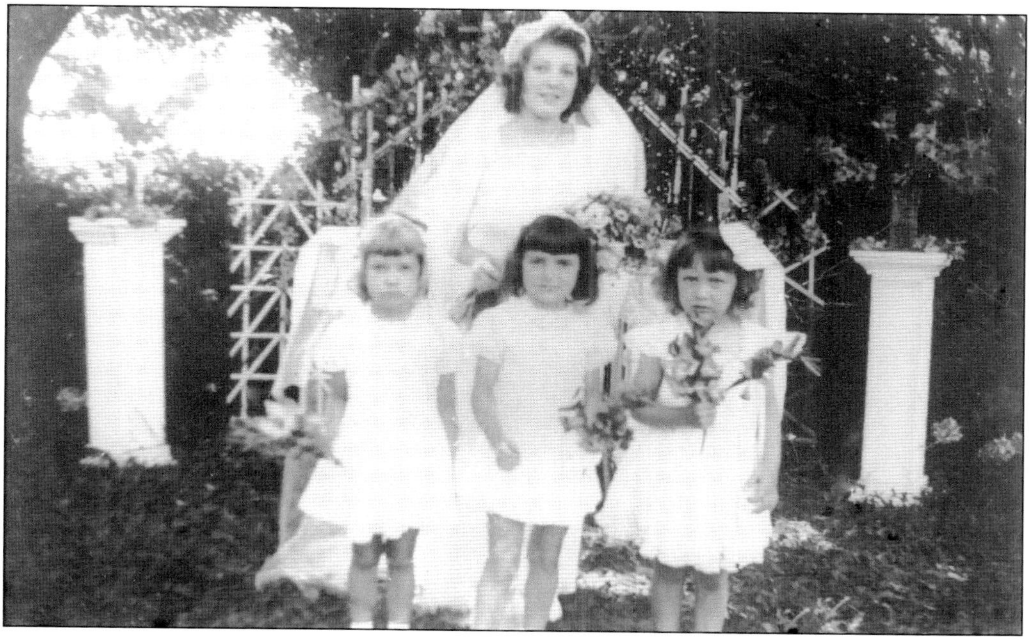

In the photograph above are three unidentified young girls and a bride on her wedding day at St. Mary's Catholic Church on Yonges Island. The exact date of photograph is not known but is believed to be in the early 1940s. Below is a photograph of an outdoor ceremony—perhaps a confirmation ceremony, as all of the children are in white—being held under the live oaks outside of St. Mary's Church. (Both, courtesy of St. Mary's Church.)

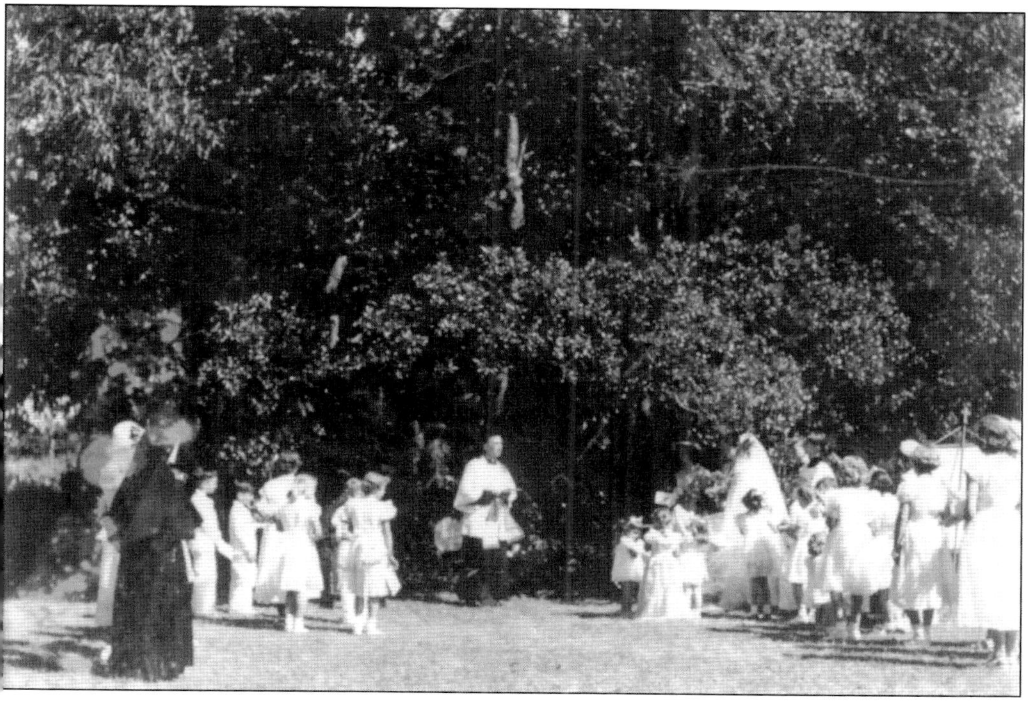

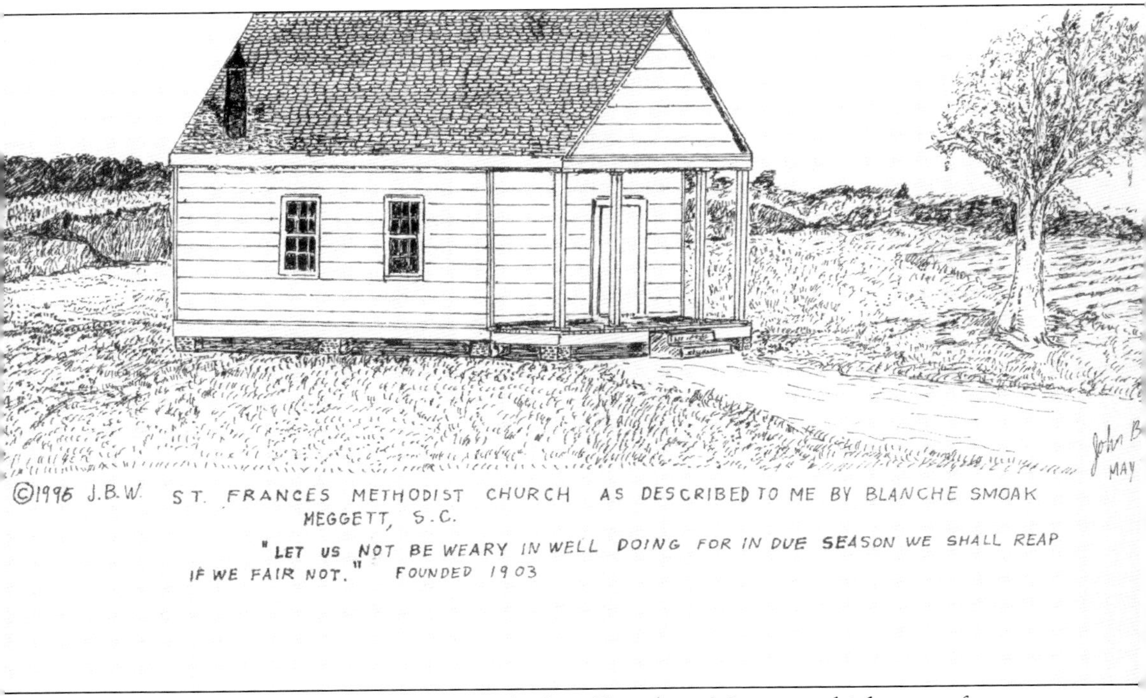

This sketch depicts St. Frances Methodist Church in Meggett, which was a frame structure located between the town of Meggett and the bridge over to Yonges Island at one point in time. The church was established in 1903 but caught fire and burned down one winter morning during a church service in the early 1940s. (Courtesy of John Waring.)

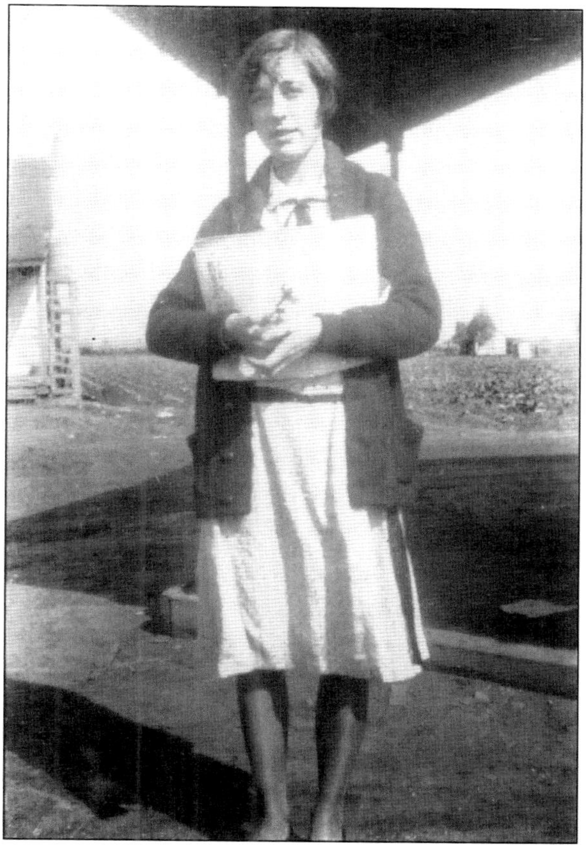

Lavonia Smoak Reaves is pictured in 1926 or 1927 when she was about 18 or 19 years old. Carrying her music books, she is on her way from her home place on Goshen Plantation to take piano lessons at Mrs. Clyde Gibson's home on Yonges Island. Reaves played the piano for St. Francis Methodist Church until it burned down. (Courtesy of Anne Commins.)

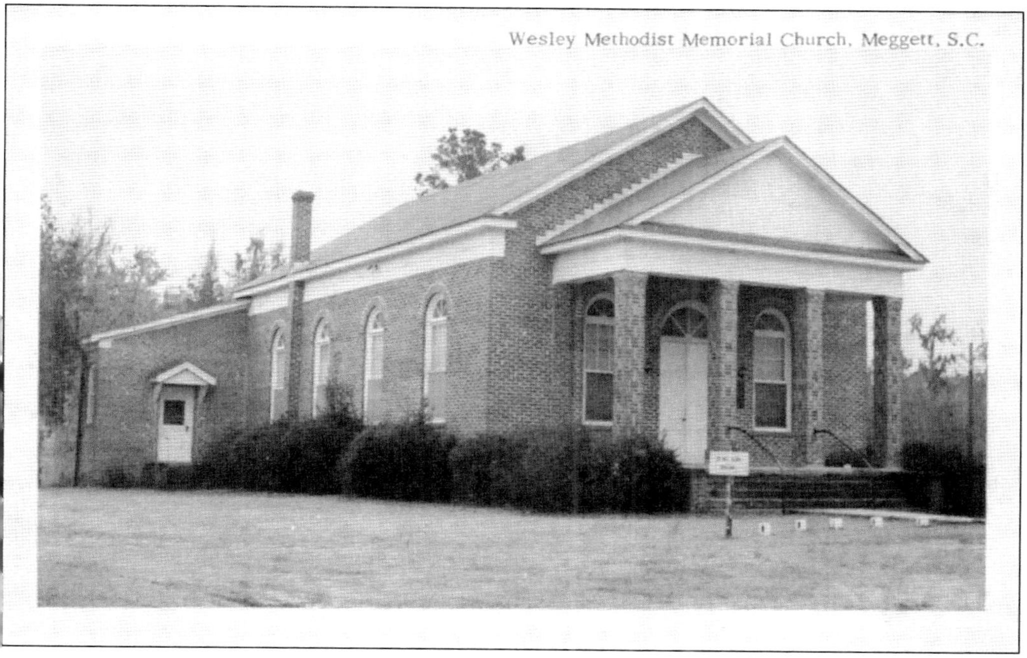

This postcard shows Wesley Memorial United Methodist Church on Highway 165 outside of Hollywood. The church was located on Highway 162 before it burned in 1985. (Courtesy of John Waring.)

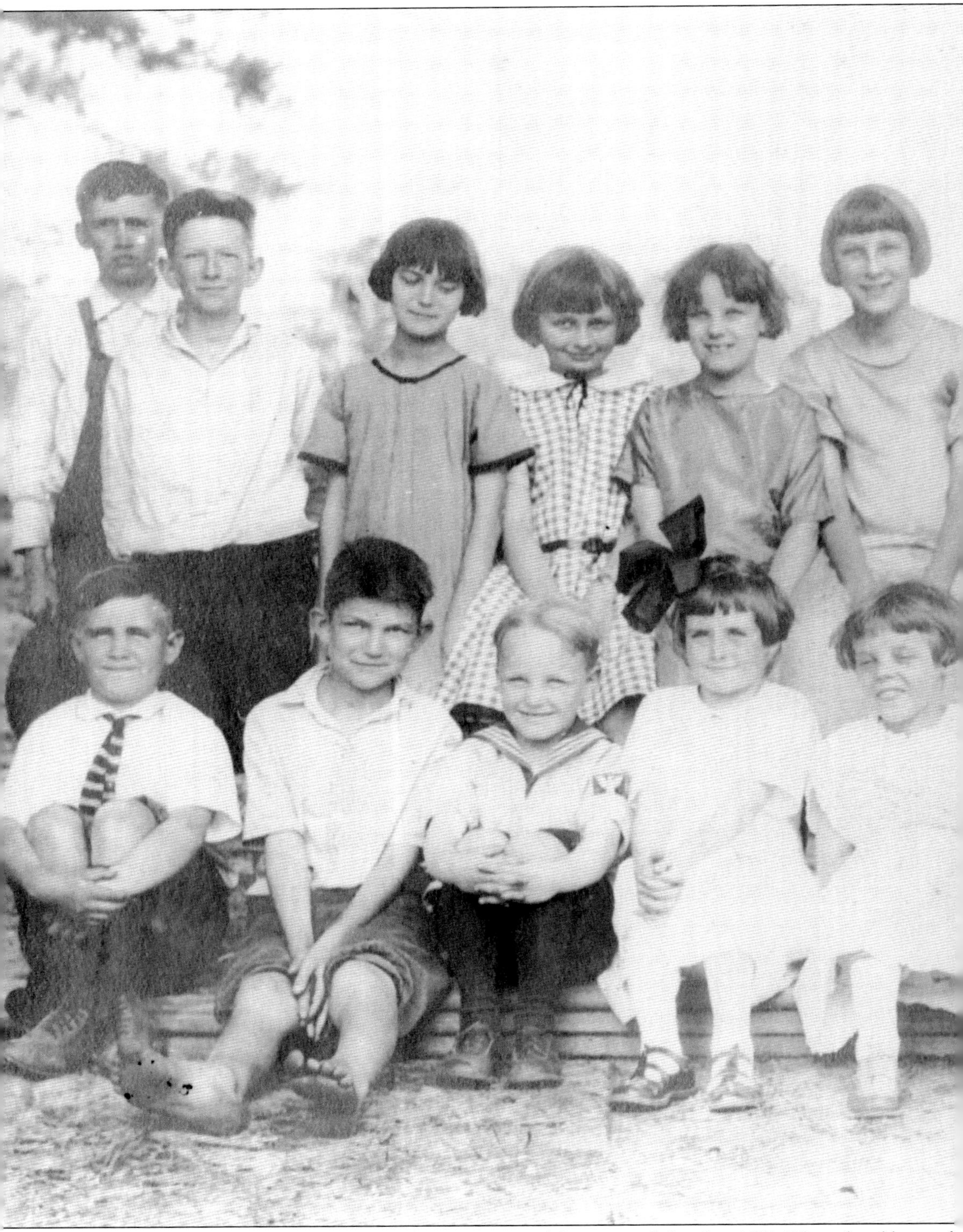

These unidentified children are students at Ravenel Grammar School in 1921. The school housed

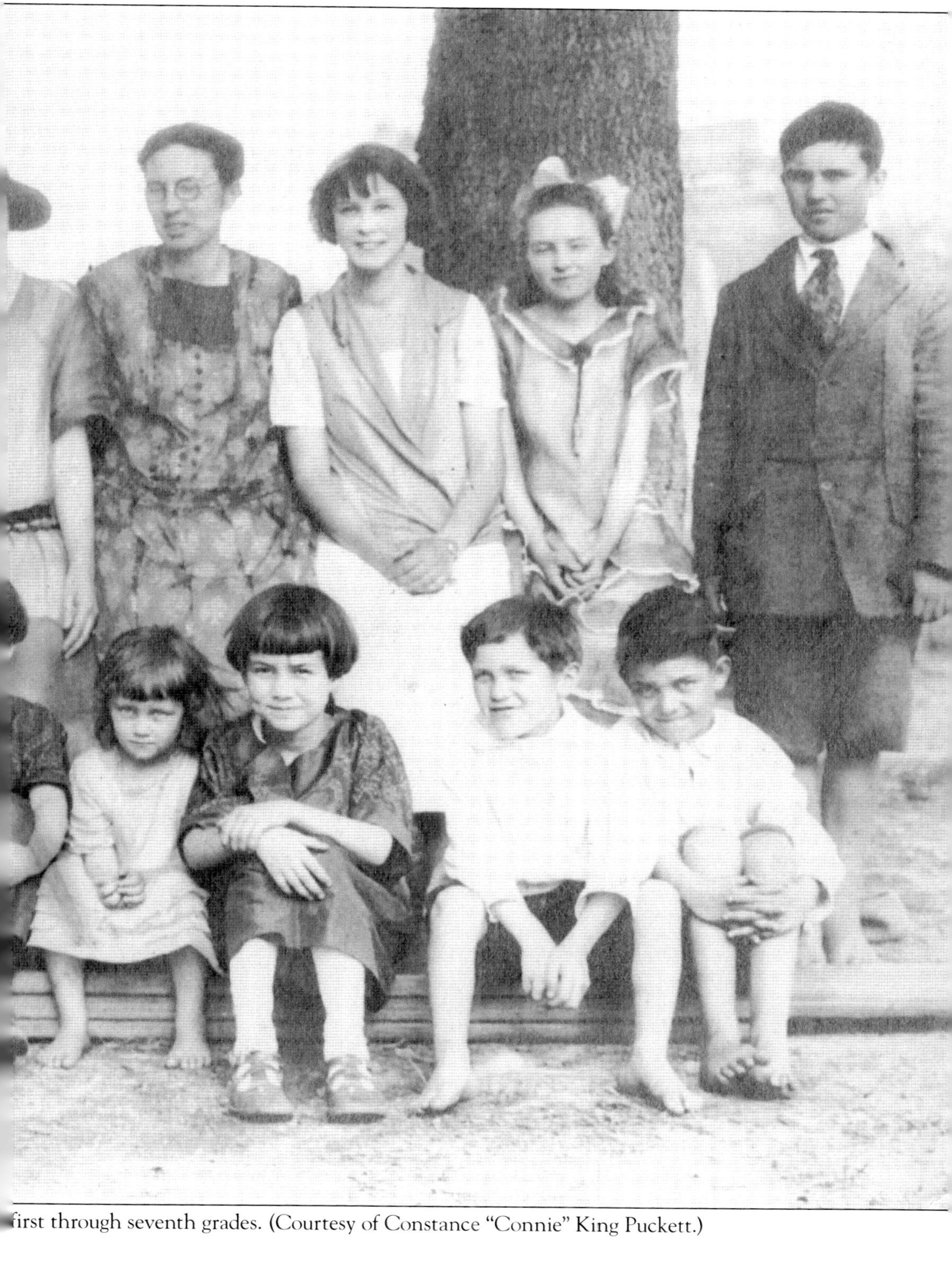

first through seventh grades. (Courtesy of Constance "Connie" King Puckett.)

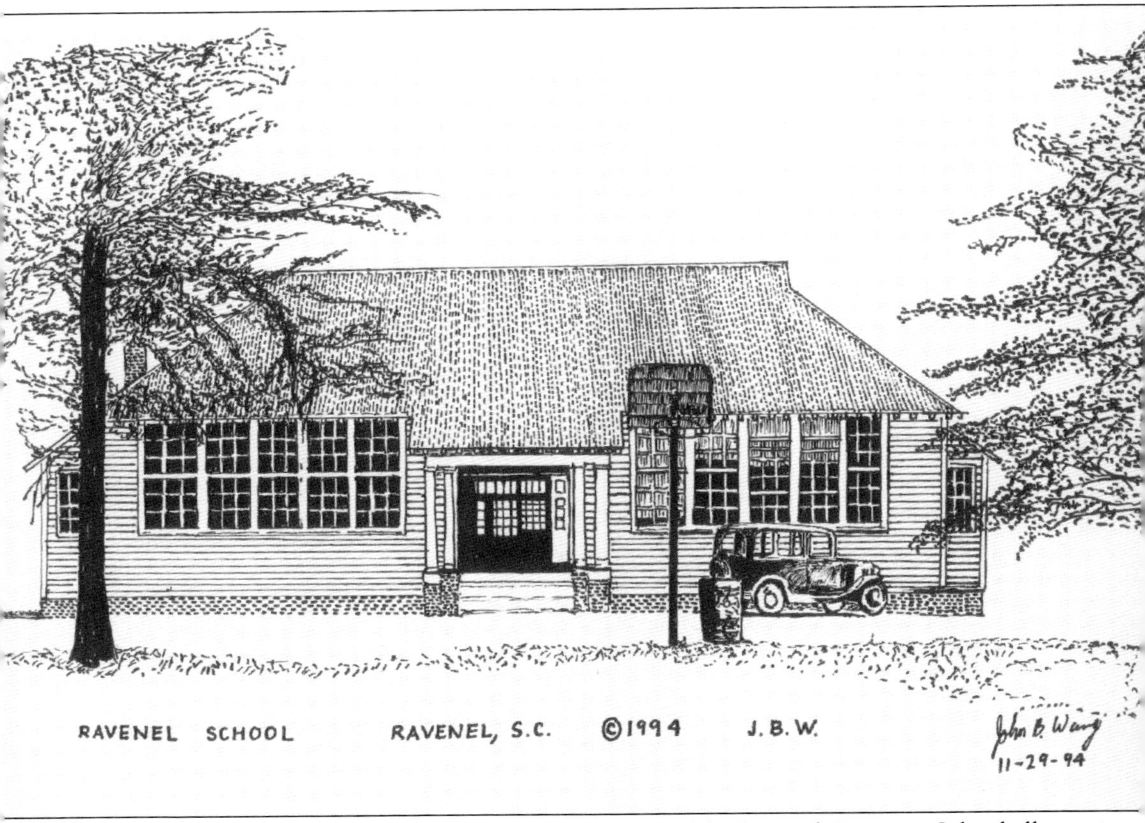

This is an artist's rendering from a historic photograph of the Ravenel Grammar School, illustrating the exterior of the one-story frame structure that served the schoolchildren of the area. (Courtesy of John Waring.)

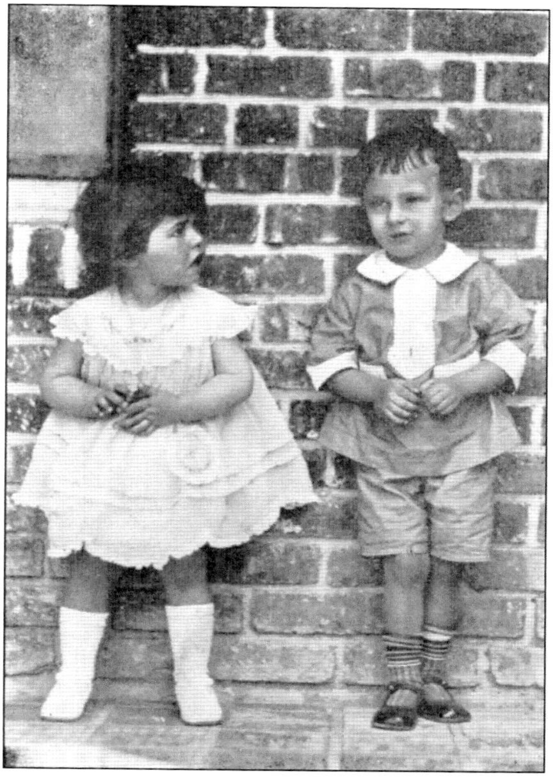

At right are two children, simply labeled "school mascots" in the Meggett School's 1926 yearbook. At the time, the Meggett School (formerly known as the Yonges Island Consolidated School) served children from first grade through high school. It was constructed in the early 1920s. According to *A Historical Sketch of Meggett Public School*, it was considered a facility "with all modern conveniences and appliances and will cost when completed $55,000." (Both, courtesy of Mona Butler Williams.)

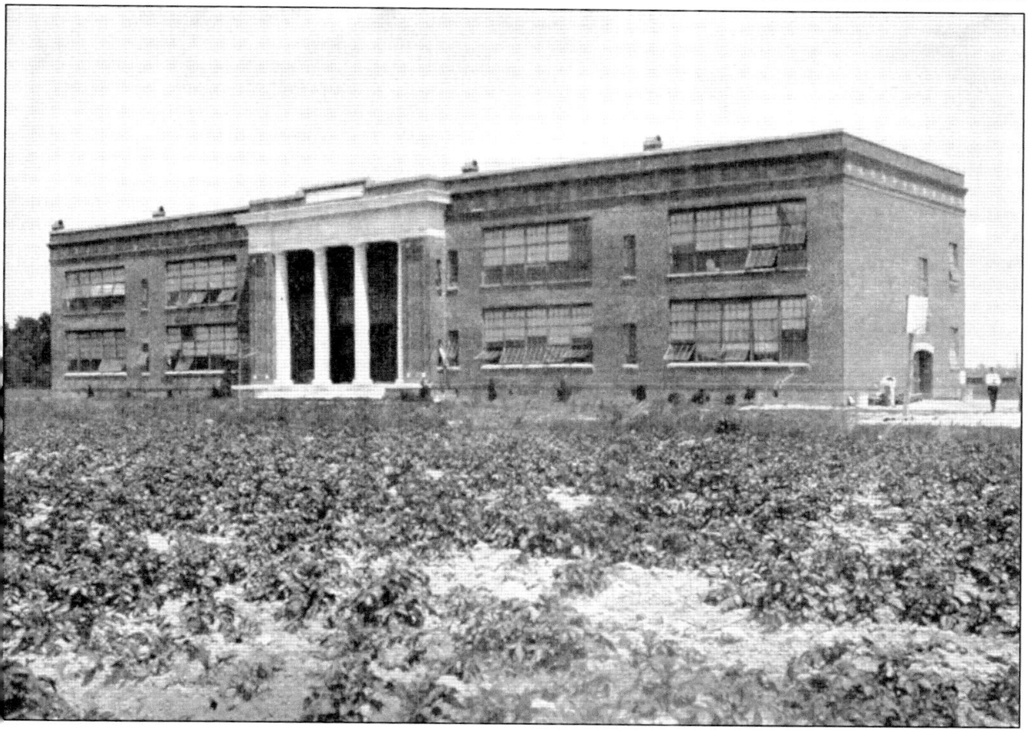

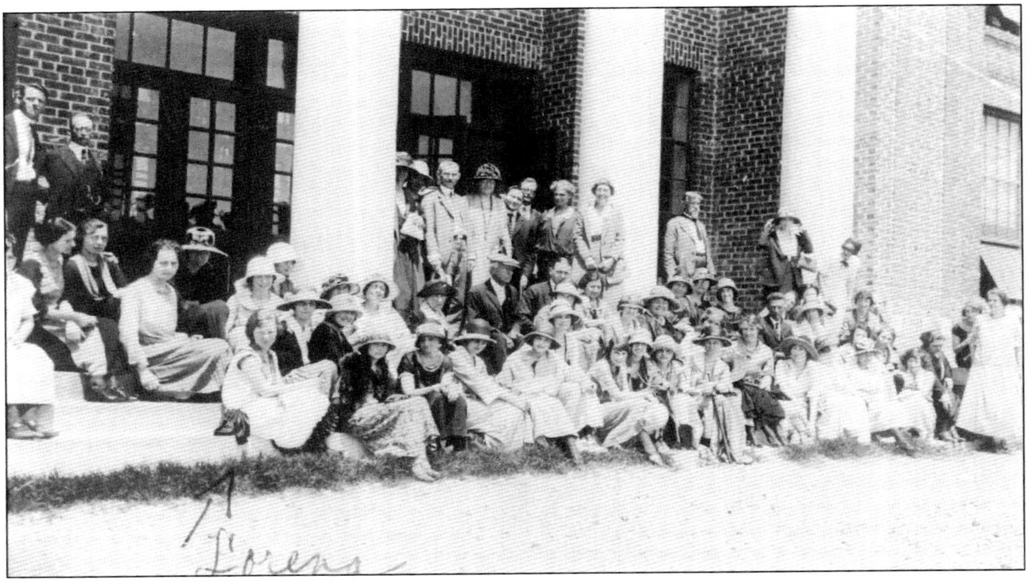

In the photograph above is the Meggett School faculty sitting outside the newly constructed school in 1923. The picture features writing from Lorena Lifrage Knisley, an English teacher at Meggett until her marriage to Charles R. Knisley, whom she met while living at Gay's boardinghouse in Meggett. Knisley went on to teach at St. Paul's High School as well many years later. Below is a lower-grade class sitting outside the school in 1925. (Above, courtesy of John Waring; below, courtesy of Natalie Towles Whiteman and Liz Whiteman.)

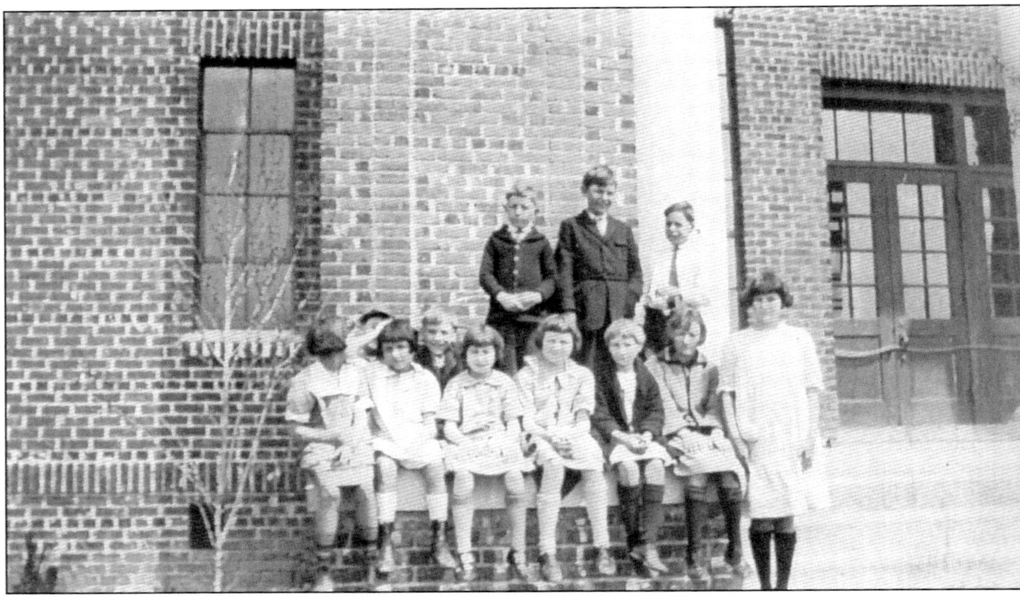

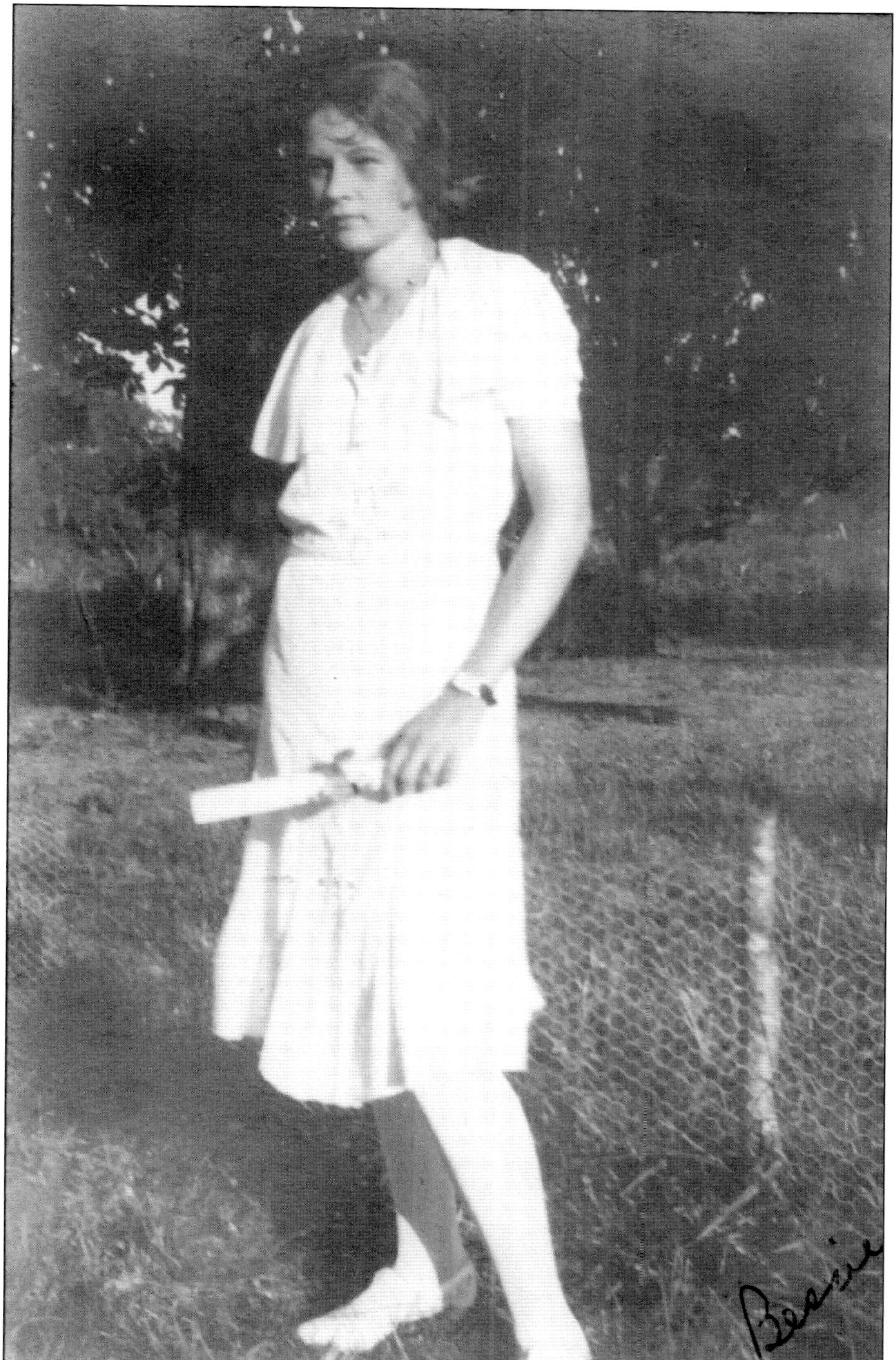

Elizabeth Roach King, also known as "Bessie," is pictured at her graduation from St. Paul's High School in the early 1930s. King was born in 1913, became the wife of Richard G. King, and was an active member in the St. Paul's Garden Club. (Courtesy of Constance "Connie" King Puckett.)

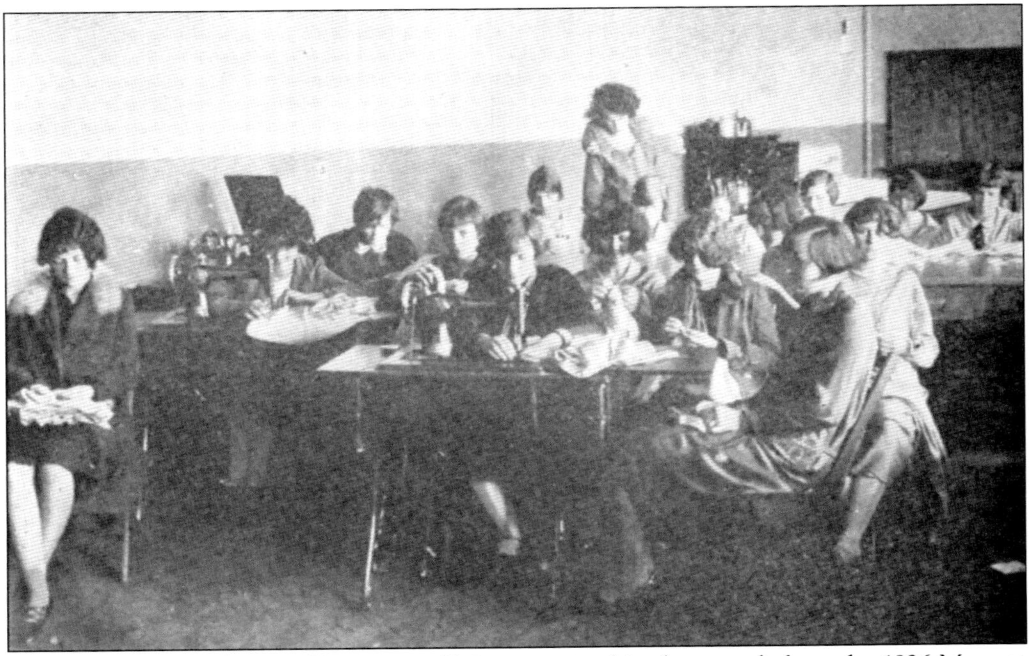

The Meggett School Home Economics Club is seen in this photograph from the 1926 Meggett yearbook. The teacher at the time was M.A. Thompson. Members of the club include, in no particular order, Lavonia Smoak, Marion Strobel, Anita Parks, Theresa Hoats, Bessie Wilson, Lucille Martin, Pearl McGill, Thelma Godfrey, Anna Baldwin, Myrtle Driggers, Margarita Chaplin, Lillian Smoak, Lelia Hethington, Laura Sauls, Lois Baldwin, and Derrill Sanders. (Courtesy of Mona Butler Williams.)

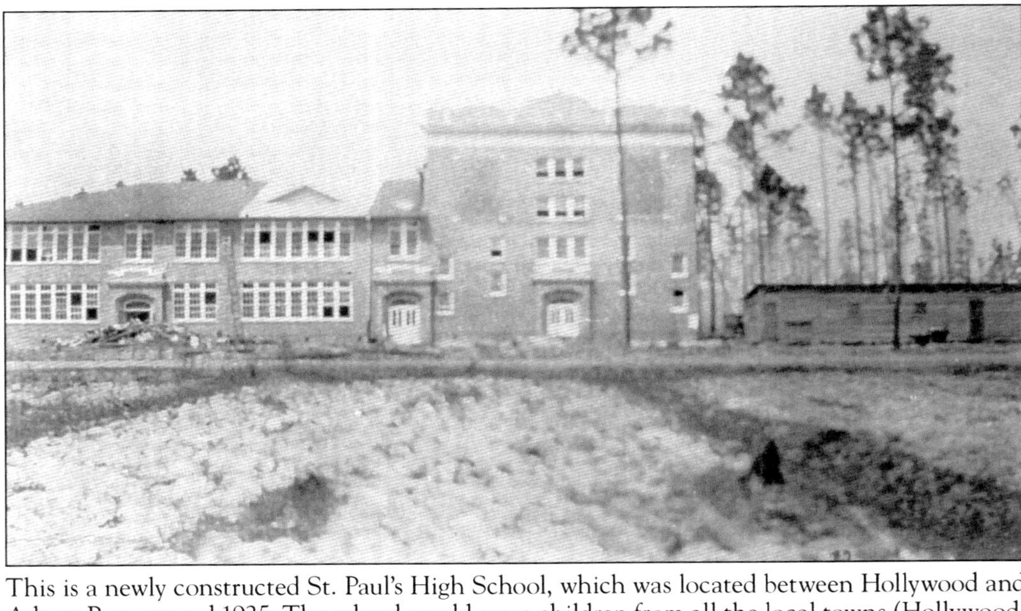

This is a newly constructed St. Paul's High School, which was located between Hollywood and Adams Run around 1925. The school would serve children from all the local towns (Hollywood, Ravenel, Adams Run, Meggett, Rantowles, and Yonges Island) for many years. (Courtesy of Natalie Towles Whiteman and Liz Whiteman.)

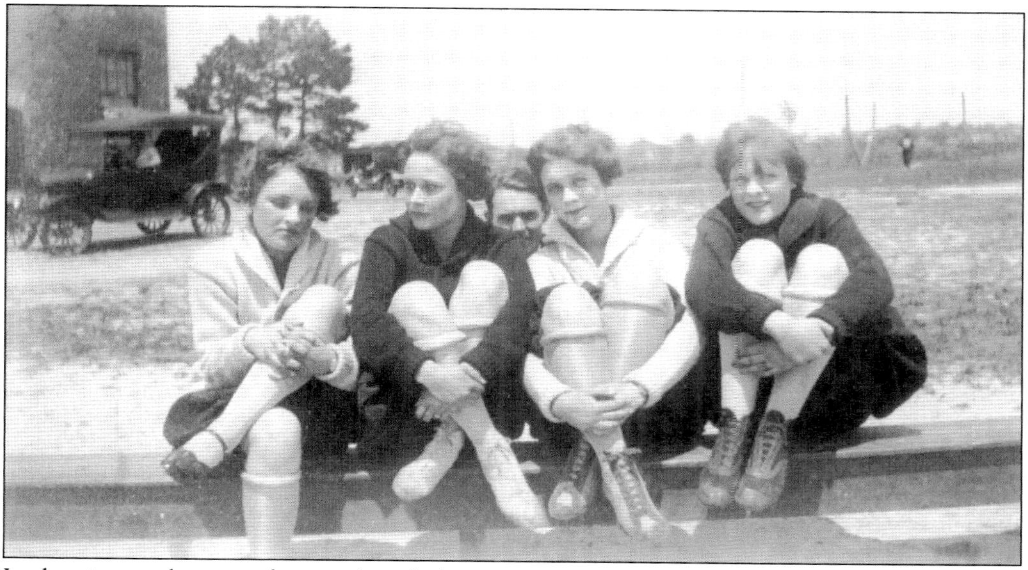

In the picture above are four unidentified members of the 1926 Meggett basketball team. The members of the team were Lois Baldwin, Mary Dodd, Marion Strobel, Anita Parks, Anna Baldwin, Bessie Wilson, and Lucille Martin. The young man in the picture is George Parish. In the photograph at right are two of the basketball players having fun outside of Meggett School one afternoon. (Both, courtesy of Anne Commins.)

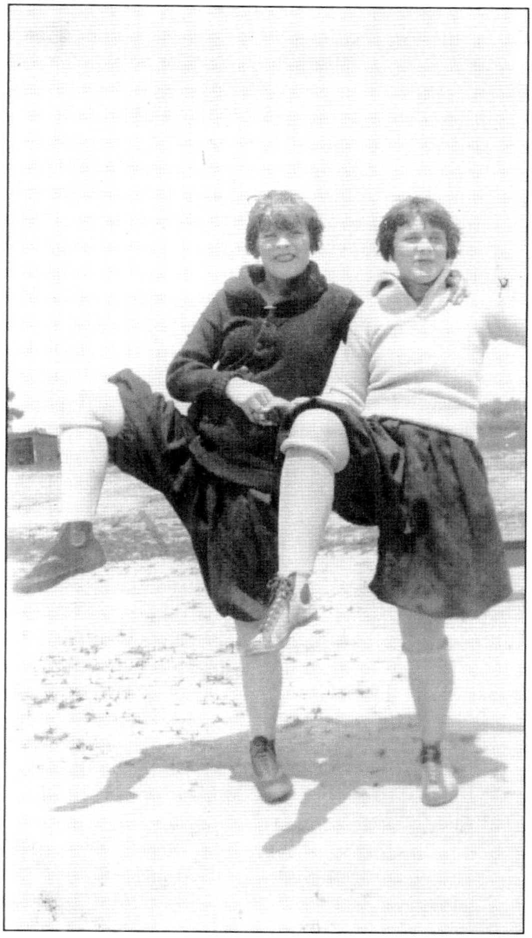

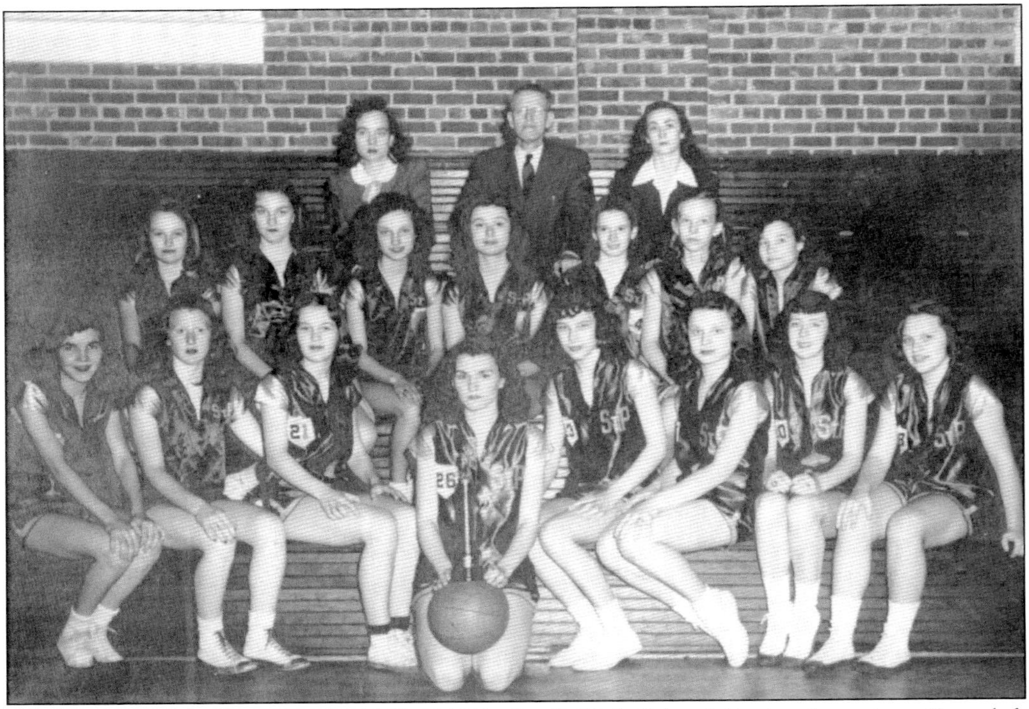

In the photograph above are the 1947–1948 St. Paul's High School district champions. From left to right are (first row) Alice Mixson, Mary Lou Knisley, Barbara Chaplin, Florence Floers, Jo Anne Towles, Patricia Youmans, LePage Thomas, and Betty Anne Welch; (second row) Claire Redmon, Betty Gay Smoak, Barbara Martin, Bea Perry, Gloria Ackerman, LaRoache Stone, and unidentified; (third row) Miss Barlett, Mr. Schroeder, and unidentified. Below is a picture of the St. Paul's High School football team in 1946. This team became South Carolina state champions. (Both, courtesy of Bobby Shealy.)

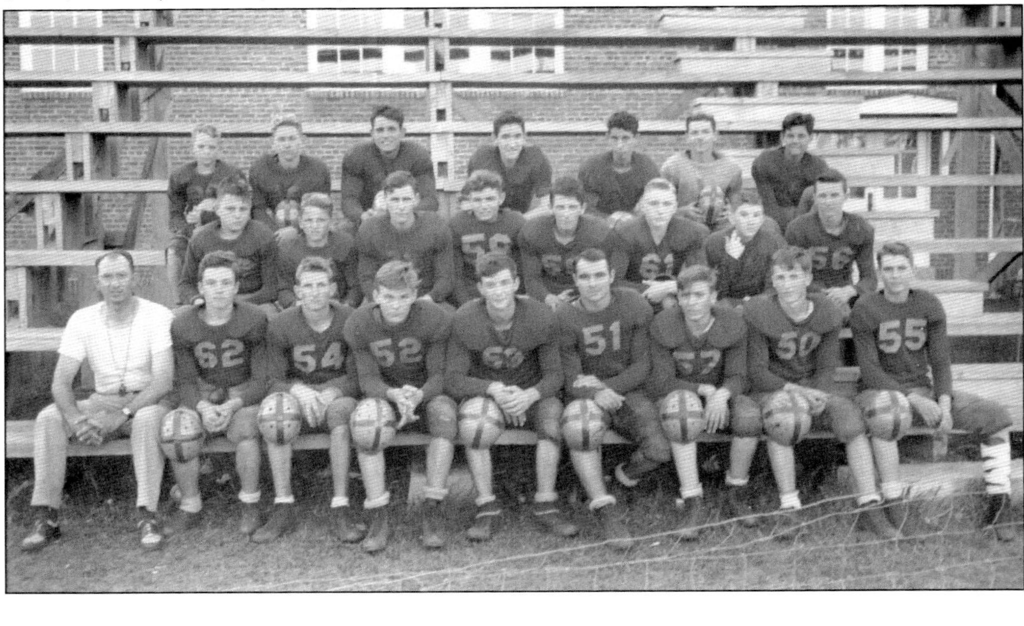

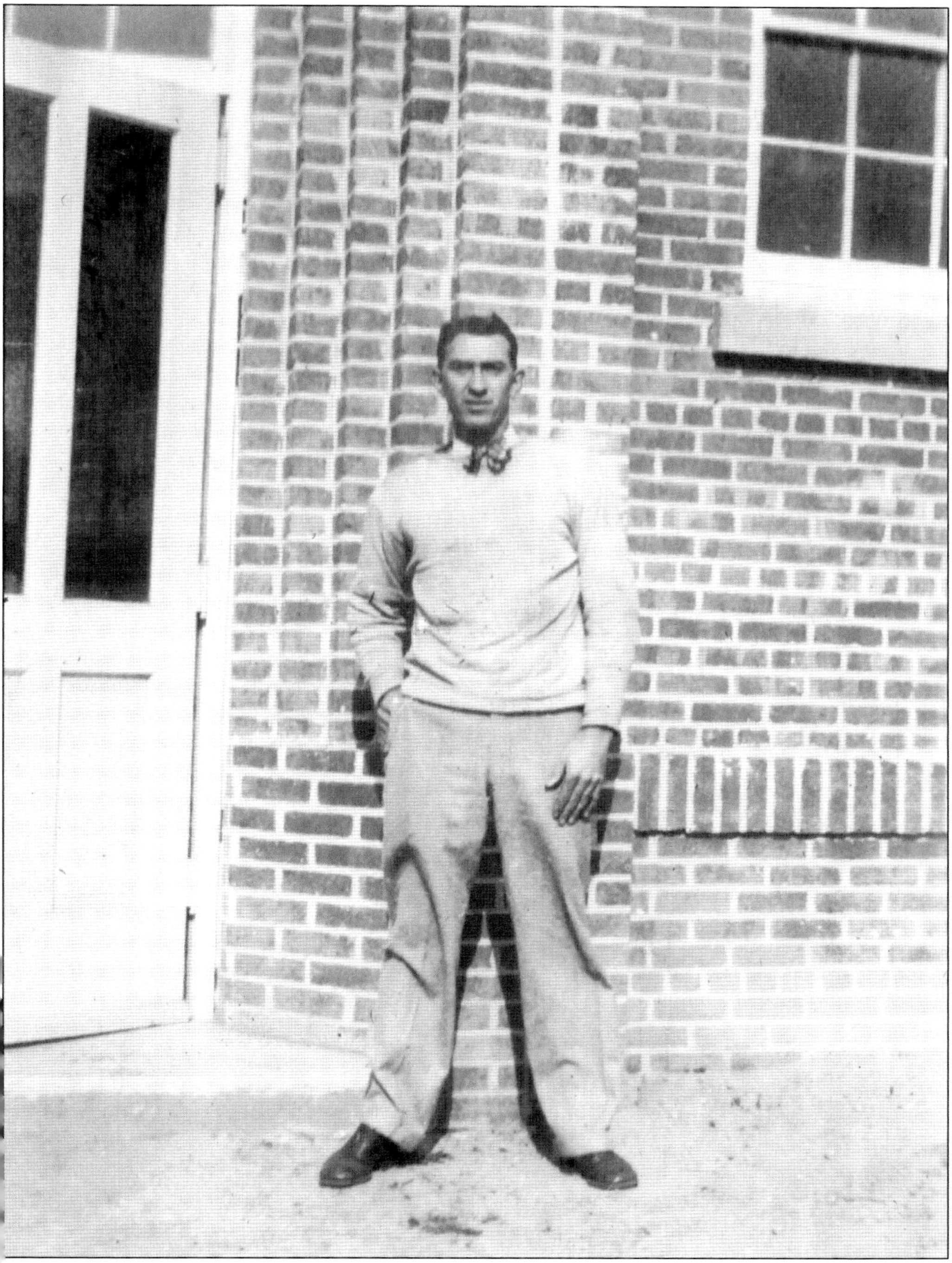

Pictured here is Charles W. Shealy, who began a 44-year career at St. Paul's High School in June 1927. Coach Shealy, as he was known, developed an athletic program that would become the envy of many high schools in the Lowcountry. Shealy was a graduate of Newberry College and taught one year at Fairfax High School in Fairfax, South Carolina, before coming and settling in at St. Paul's High School. (Courtesy of Bobby Shealy.)

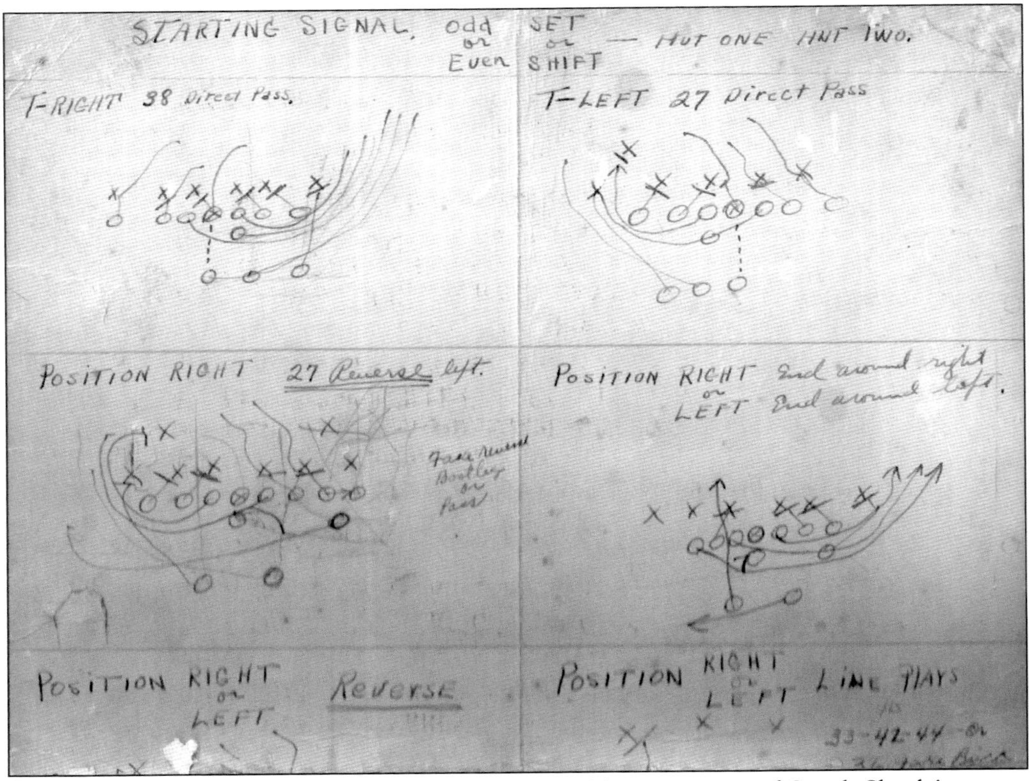

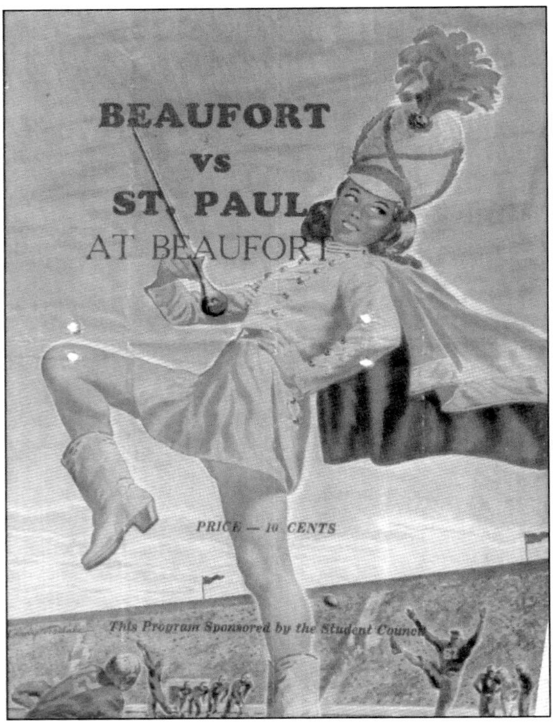

Above is a copy of Coach Shealy's playbook, written on the back of a football program similar to the one seen in the picture at left. The diagrams were written during game time to show his players what they needed to do to win. Coach Shealy was considered by many to be a mastermind when it came to calling plays on the field. During his tenure at St. Paul's, he had a most impressive record, including two state and four district championships in Class C football; four district championships in Class B football; one baseball championship in Class B baseball; eleven district and four lower-state championships in boys' basketball; and eight district, two lower-state, and one state championship in Class B girls' basketball. (Both, courtesy of Bobby Shealy.)

The pirate was the mascot of St. Paul's High School. This image was taken from a prom dance card from 1946, belonging to Bobby Shealy. St. Paul's was home of the Pirates, and the yearbook was the *Treasure-Chest*. Today, St. Paul's Academy, a private school in Hollywood, has adopted the Pirate mascot to keep the tradition going. (Courtesy of Bobby Shealy.)

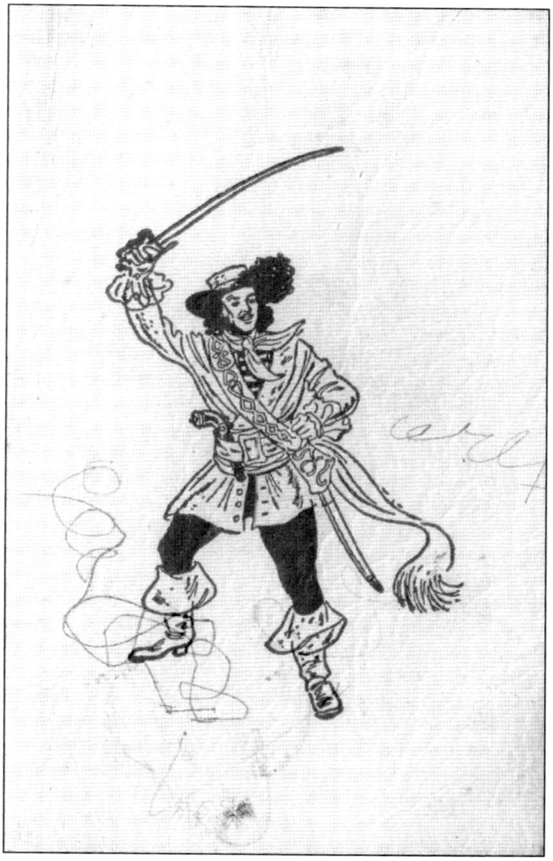

This postcard shows St. Paul's Academy, which is a private school located in Adams Run that opened in the mid-1960s. St. Paul's Academy served children through the eighth grade, while high school students attended its sister school, St. Paul's Country Day, in Hollywood. Today, the academy is located in the high school building and continues to serve students from three-year-old kindergarten through eighth grade. (Courtesy of Mona Butler Williams)

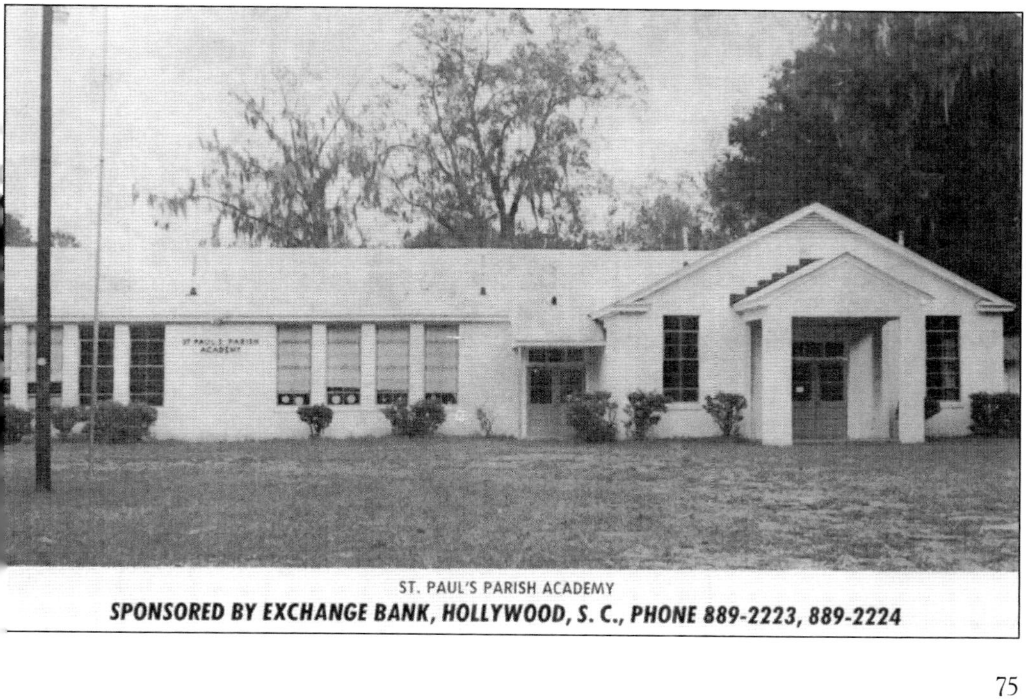

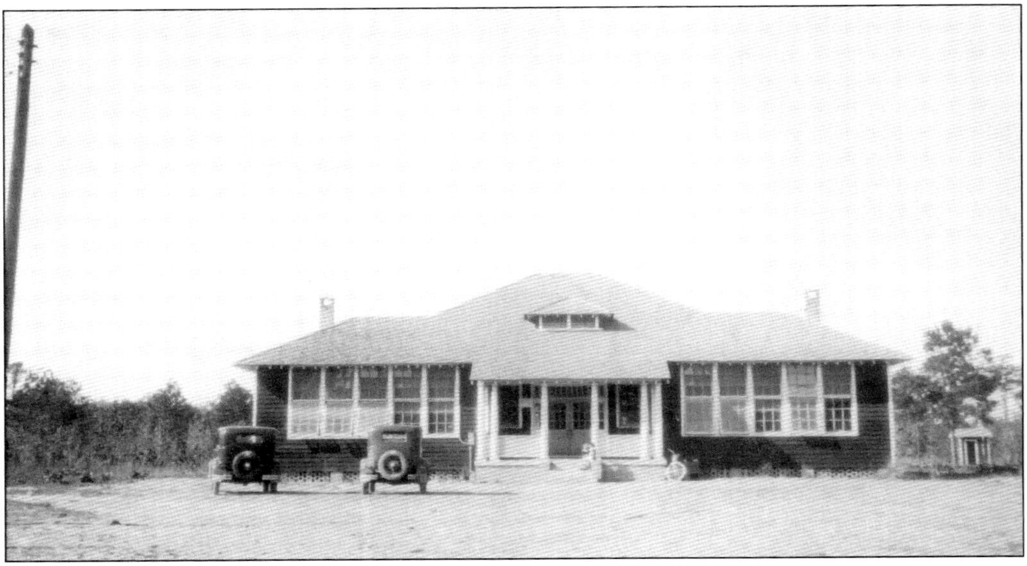

In the photograph above is Rantowles Elementary School around 1930. The school was located on Highway 162 in St. Paul's Parish and served as a gathering and learning place for the community's children. Below is a graduation invitation card for the school, belonging to Ida May Driggers. (Both, courtesy of the Charleston County School District.)

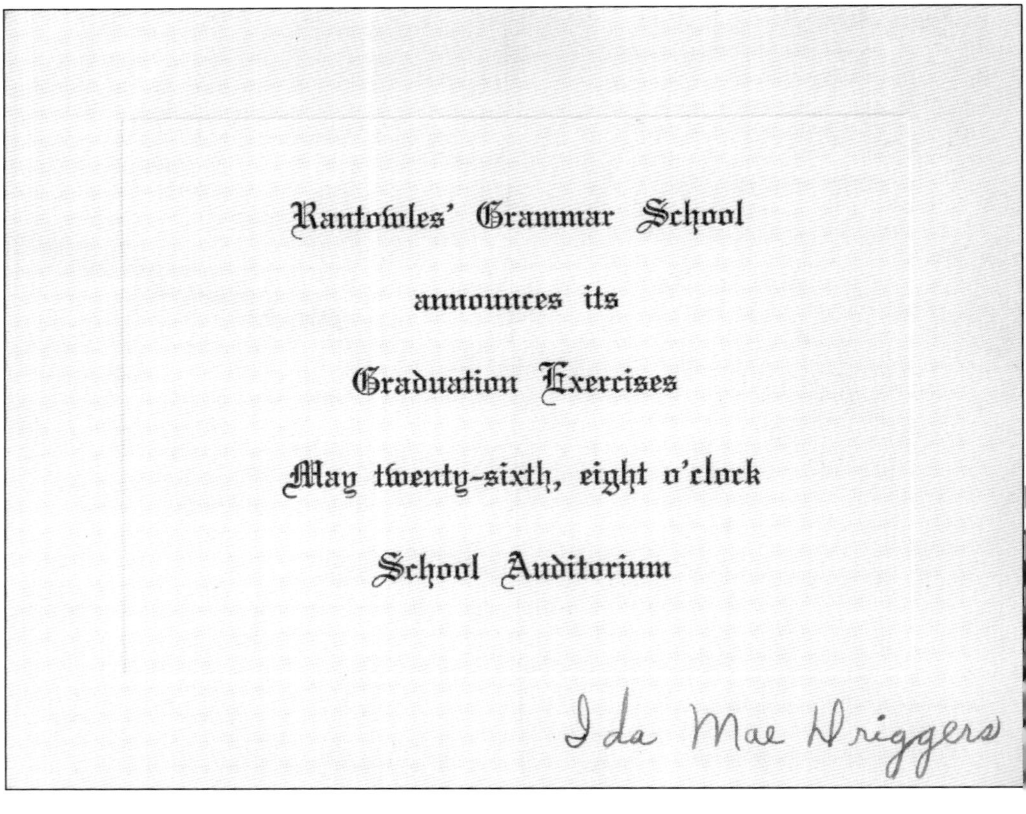

Rantowles' Grammar School

announces its

Graduation Exercises

May twenty-sixth, eight o'clock

School Auditorium

Ida Mae Driggers

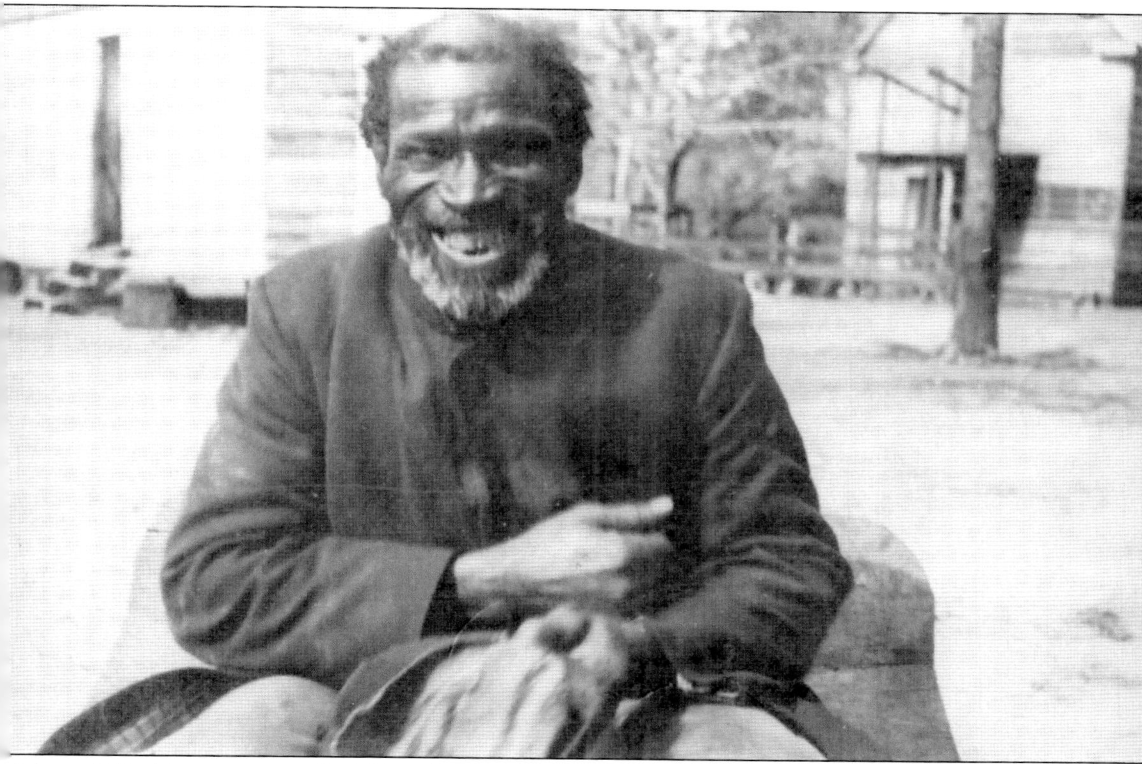

Ishmael Washington was called "Daddy Ishmael" by Adams Run schoolchildren. Washington worked for Dr. Tally Taylor, a local physician, but he also was the custodian for the Adams Run Grammar School and a beloved figure on the campus. (Courtesy of Anna Blythe.)

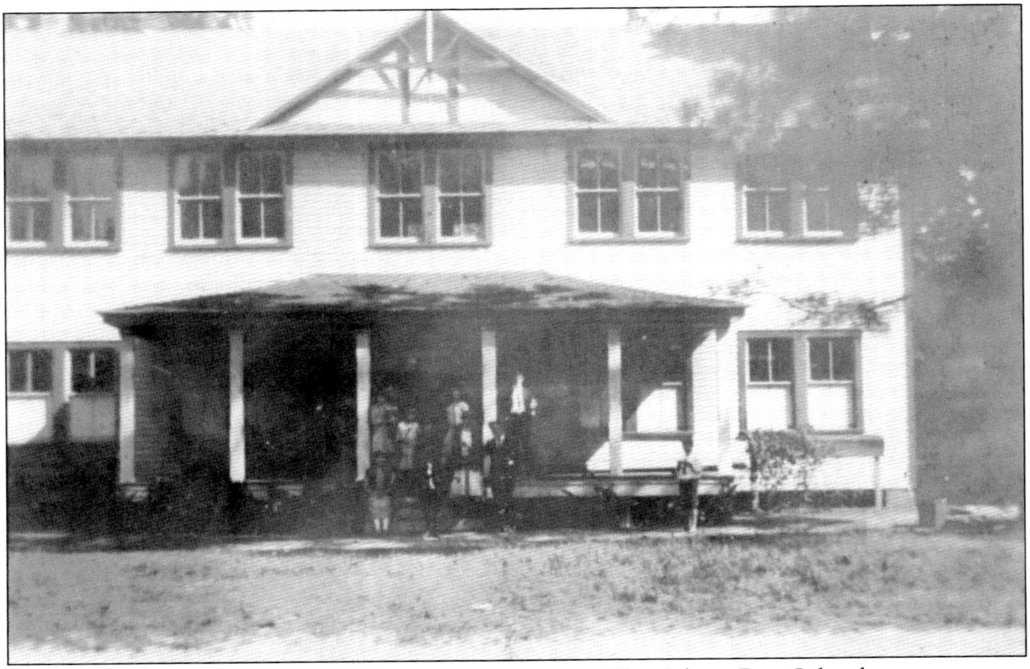

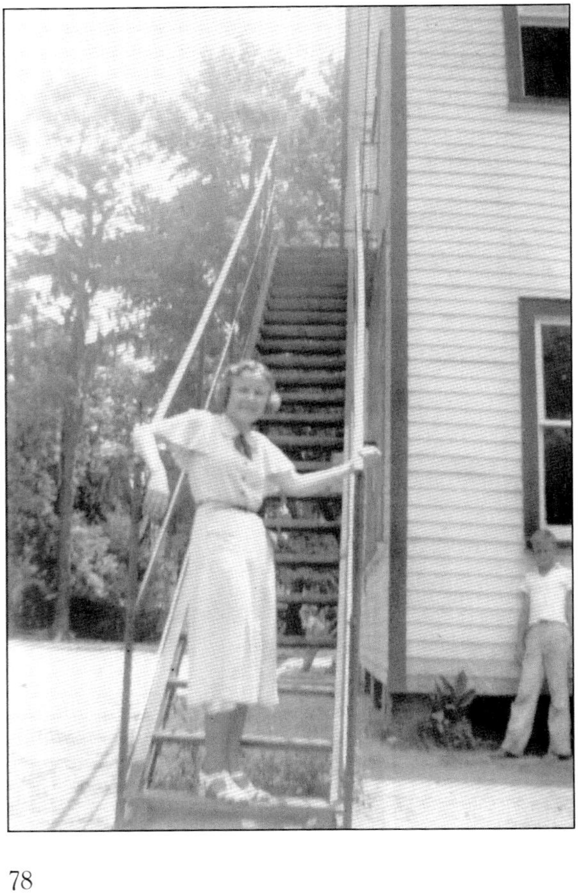

The Adams Run School was a significant two-story, frame building located in the heart of the Adams Run village. It served children from the surrounding area in the first through eighth grades. The structure is no longer standing. (Courtesy of Jackie Boineau.)

Thelma Rankin is standing on the fire escape outside of Adams Run Grammar School in the early 1930s. Rankin was the first- and second-grade teacher. She lived in Charleston but drove out daily to attend to her students. (Courtesy of Jackie Boineau.)

Three

TRAINS, PLANES, AND AUTOMOBILES

Methods of transportation have always been and will always be an important aspect of American society. Whether by trains, planes, or automobiles, people and goods are moved from one place to the next, and St. Paul's Parish is no different. Here, a steam engine is sitting at the Ravenel train depot around 1920. (Courtesy of Constance "Connie" King Puckett.)

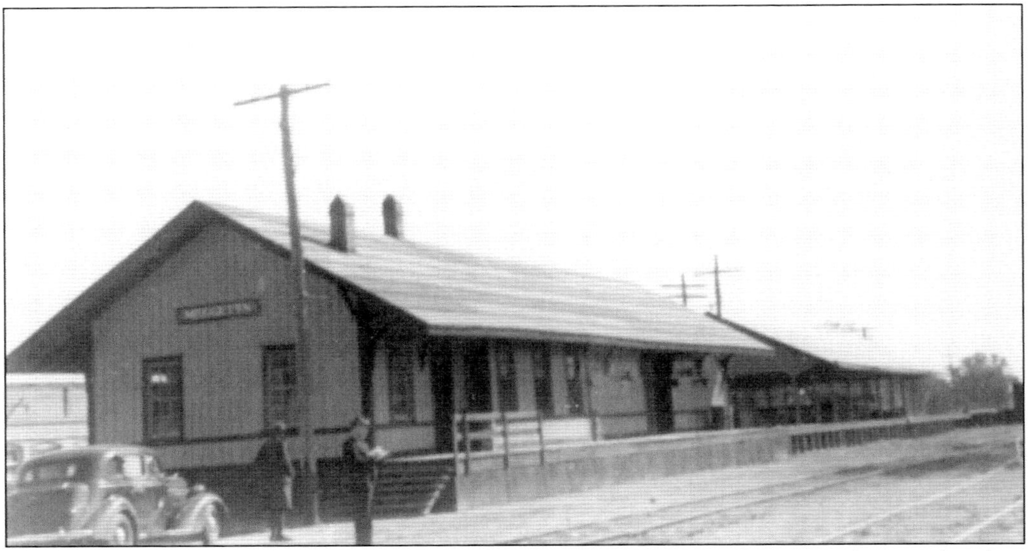

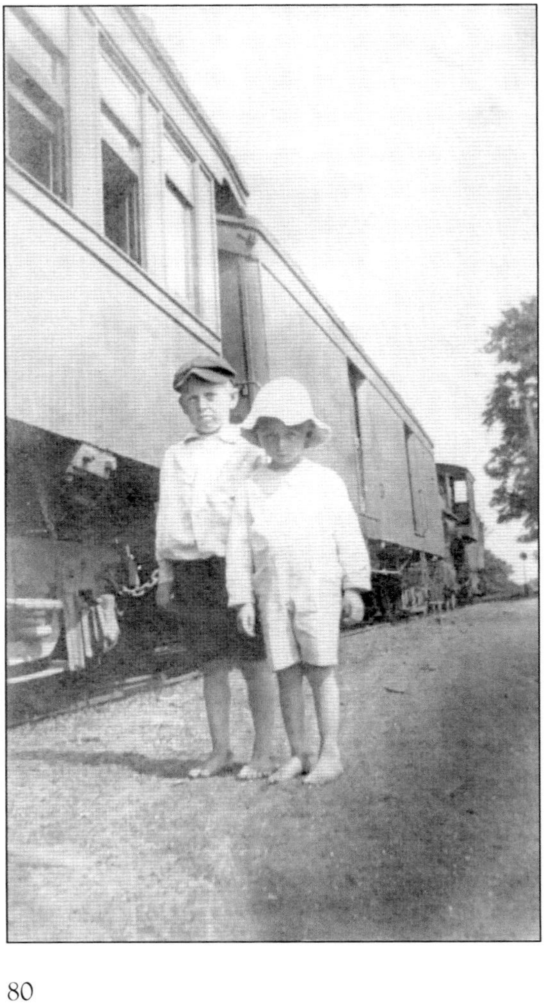

This is the Atlantic Coastline Railroad Depot in Meggett in the early 1940s. The Atlantic Coastline was an American railway system that was in existence between 1900 and 1967 and had a significant presence in the Southeast. The Meggett Depot was considered a hub for both passenger and freight transportation and was torn down in the 1970s. (Courtesy of Buddy Hill.)

Shown standing in front of the train in Meggett in 1919, these first cousins are Edwin Hoertz (left), son of the maritime painter Frederick J. Hoertz, and John Griner, who was six years old at the time. Edwin had come from up north to visit his extended family for the summer. (Courtesy of Anne Sanders Groover.)

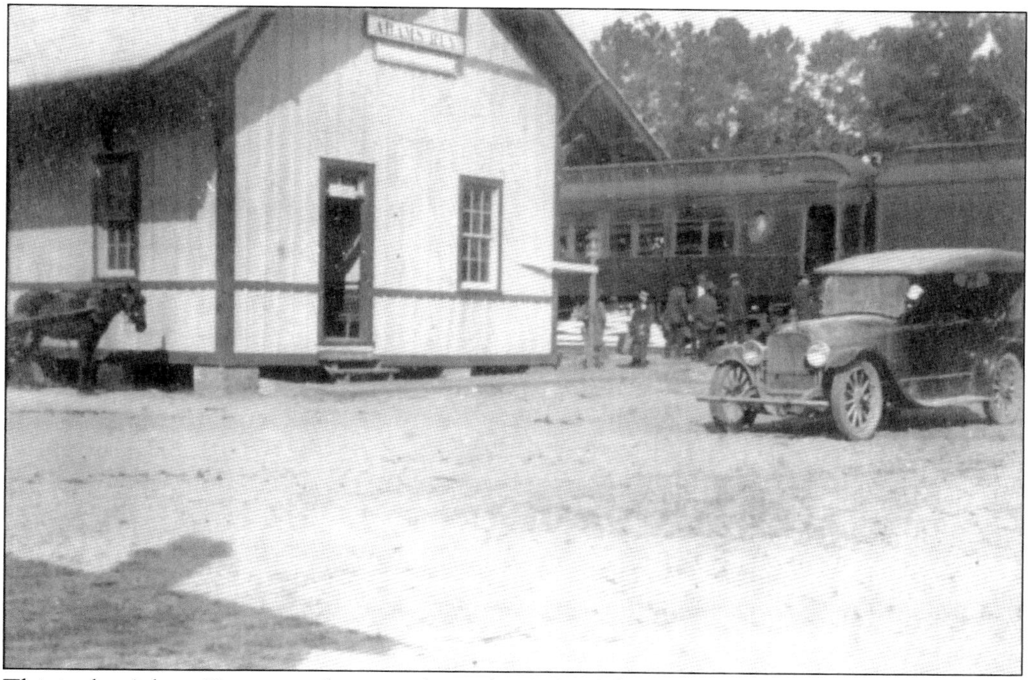

This is the Adams Run train depot in the early 1940s. Adams Run is located roughly four miles from Hollywood, along Highway 162. The depot was typical of most small-town train stations in America at that time and served as a way station for passenger and freight transportation to and from the larger cities of Savannah and Charleston. (Courtesy of Jackie Boineau.)

Pictured are, from left to right, Gordon Kears, who worked at the Rose Hill Mercantile; Jack Rumph (Kears's boss), who owned and ran the Rose Hill Mercantile, was a large-scale farmer, and owned the barrel factory; Hazel Boineau, a local railroad agent and brother of Carroll; Gus Gatch, section master for the railroad and master mechanic; and Carroll and Harry Boineau. They met at the train depot one Sunday morning to get the paper. (Courtesy of Jackie Boineau.)

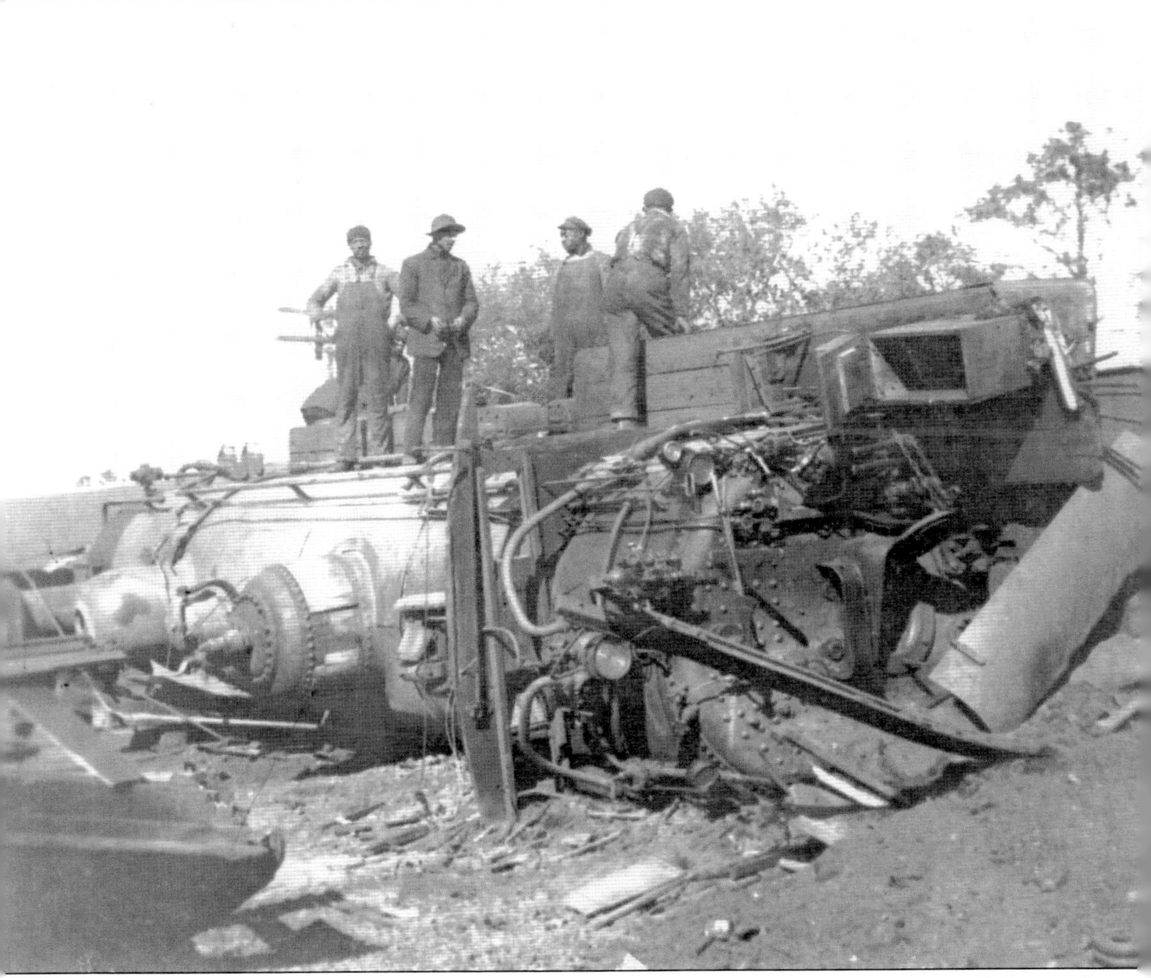

Atlantic Coastline Train No. 85, known as the "West Indian Fast Mail," jumped the tracks at Ravenel one morning while traveling an estimated 40 miles per hour. The locomotive was badly damaged, but all personnel escaped without serious injury. (Courtesy of Buddy Hill.)

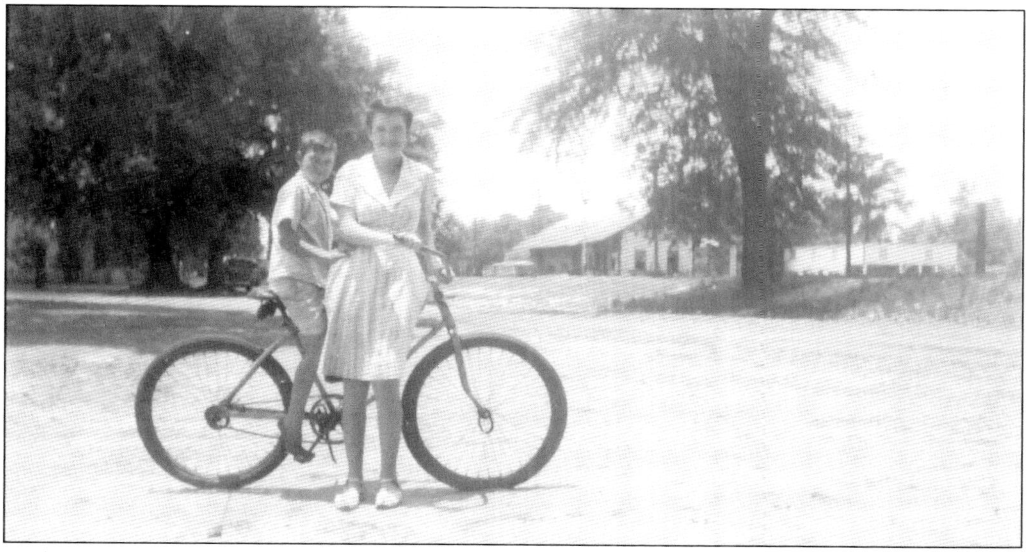

In the photograph are two unidentified Ravenel residents, perhaps mother and son, out riding a bike in the early 1940s. In the background are the Ravenel depot as well as the adjacent Hooker's Store. (Courtesy of Buddy Hill.)

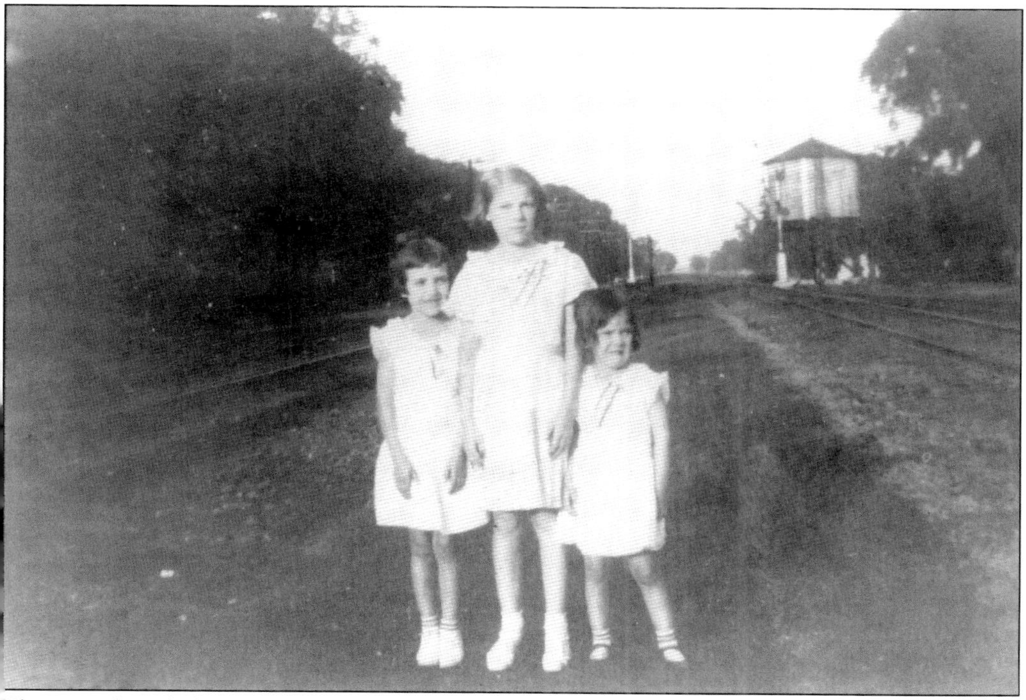

Three unidentified young ladies pose for a picture in what appears to be their Sunday best in front of the water tower located next to the railroad bed in Ravenel. The water tower was located down the tracks from the Ravenel depot and is no longer standing today. (Courtesy of Buddy Hill.)

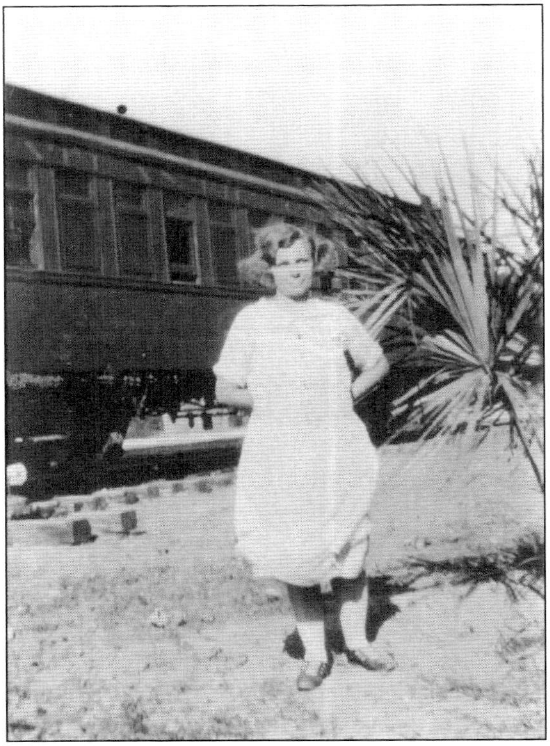

In this photograph believed to have been taken in the early 1920s, an unidentified woman is standing in front of a train that has come to a stop outside of Ravenel. (Courtesy of Buddy Hill.)

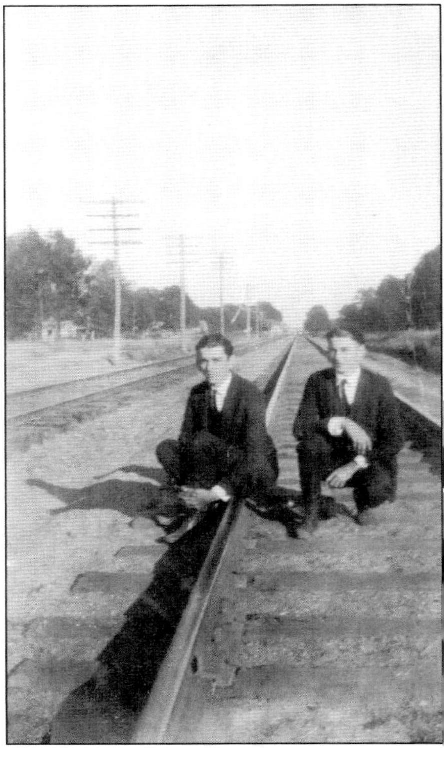

Local boys Carl (left) and Frank Butler pose for a picture while sitting on the railroad tracks in Ravenel in 1923. (Courtesy of Mona Butler Williams.)

The unidentified family in the picture at right is thought to have rented a home near the Ravenel depot at some point in the early 1940s. The timber business was booming in St. Paul's Parish in the early 20th century as local businessmen ran lumber operations in conjunction with farming. The railroads had encouraged the extraction of trees over the years to make way for new tracks, and this practice subsequently allowed for additional lands to be planted by local farmers for produce. In the photograph below is a steam engine that operated south of Adams Run on one of the numerous logging railroads in the area. A topographic map from 1918 suggests that this track was owned and operated by the Hollywood Manufacturing Company. (Right, courtesy of Buddy Hill; below, courtesy of Anna Blythe.)

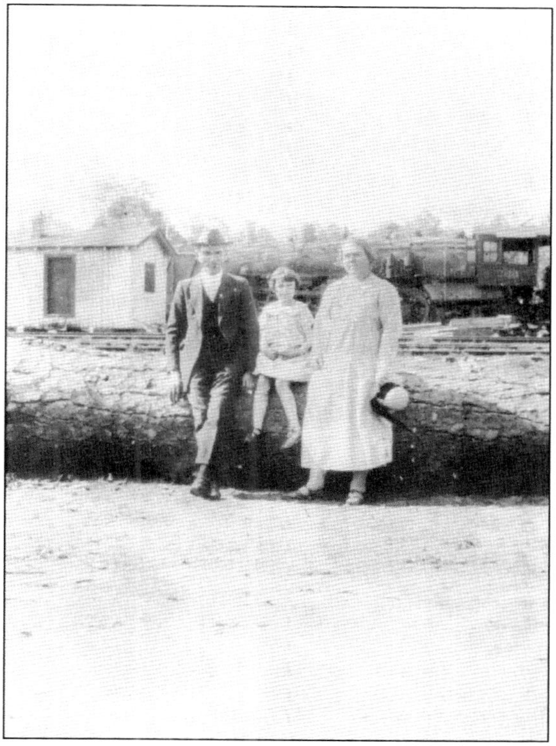

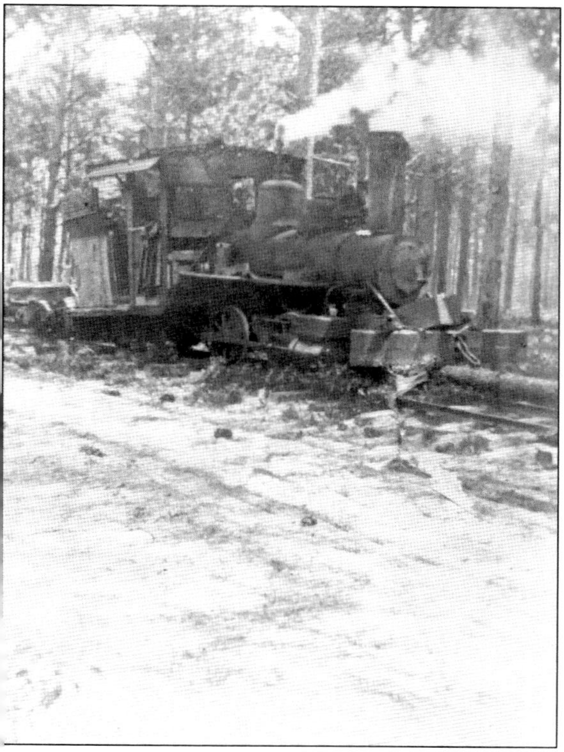

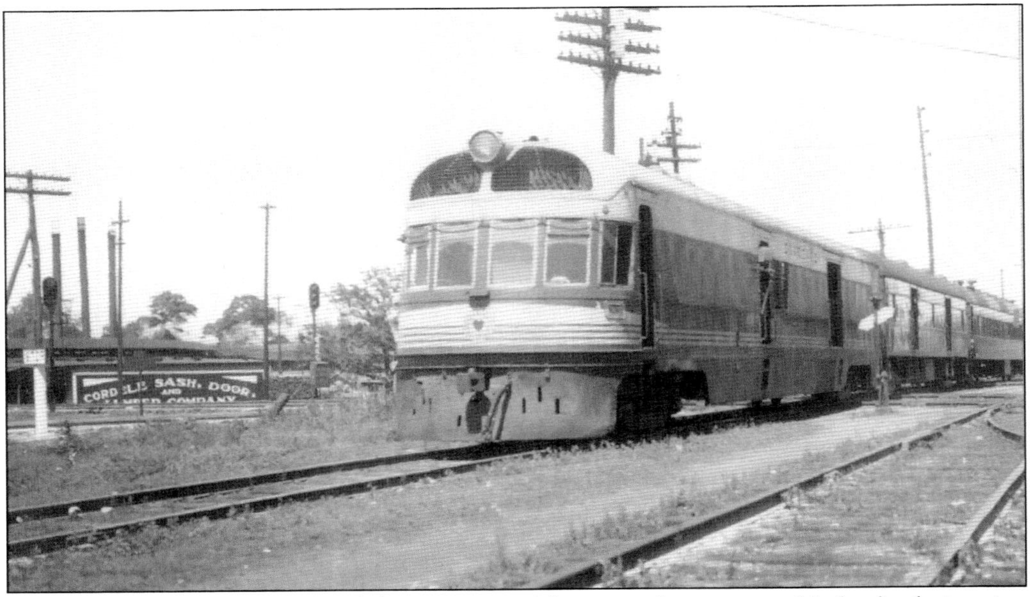

The boll weevil was a beetle that fed on cotton and flowers and was responsible for the decimation of cotton crops throughout the Southeast in the late 19th century and at other times during the early 20th century. The nickname "boll weevil" was used by locals to describe the Seaboard passenger train that ran through the parish because it was painted a dark green and looked like a boll weevil. (Courtesy of Buddy Hill.)

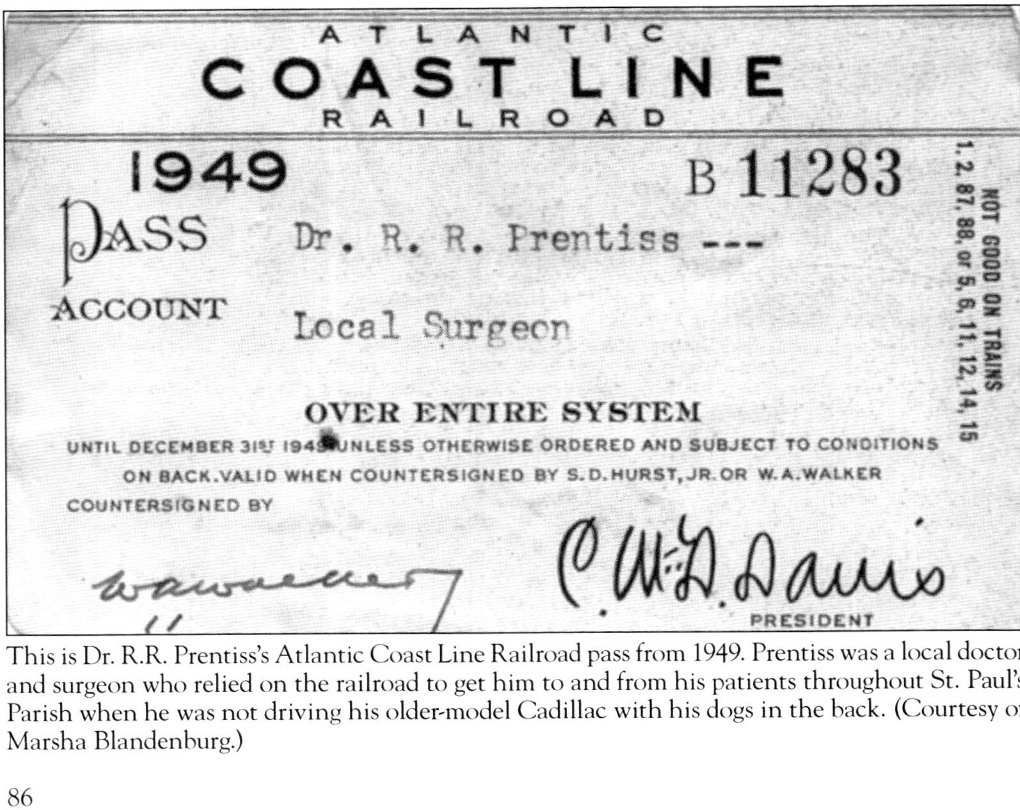

ATLANTIC
COAST LINE
RAILROAD

1949 B 11283

PASS Dr. R. R. Prentiss ---

ACCOUNT Local Surgeon

NOT GOOD ON TRAINS 1, 2, 87, 88, or 5, 6, 11, 12, 14, 15

OVER ENTIRE SYSTEM

UNTIL DECEMBER 31ST 1949 UNLESS OTHERWISE ORDERED AND SUBJECT TO CONDITIONS ON BACK. VALID WHEN COUNTERSIGNED BY S. D. HURST, JR. OR W. A. WALKER

COUNTERSIGNED BY

W. A. Walker

C. McD. Davis
PRESIDENT

This is Dr. R.R. Prentiss's Atlantic Coast Line Railroad pass from 1949. Prentiss was a local doctor and surgeon who relied on the railroad to get him to and from his patients throughout St. Paul's Parish when he was not driving his older-model Cadillac with his dogs in the back. (Courtesy of Marsha Blandenburg.)

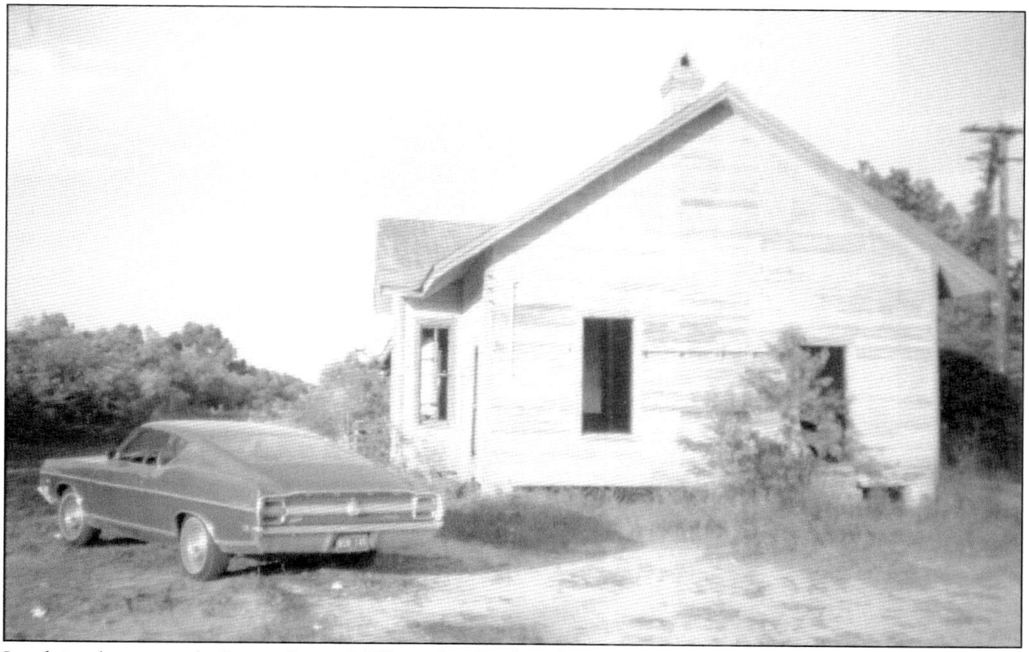

In this photograph from about 1972 is the Seabord Airline Depot in Meggett. The Seaboard Airline Railroad was a competitor to the Atlantic Coastline Railroad until they merged on July 1, 1967, and later went on to become CSX. Before it was torn down in the late 1970s, the depot was located near the intersection of current-day Highway 165 and Toogoodoo Road. (Courtesy of John Waring and Buddy Hill.)

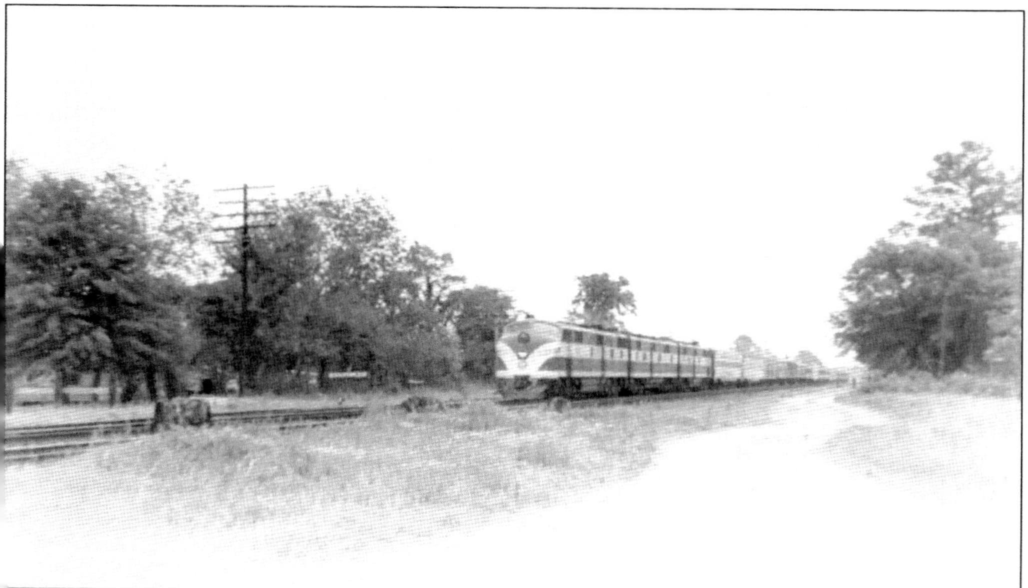

In this R.W. Cartwright photograph, a streamliner train known as the Champion is entering the community of Rantowles in St. Paul's Parish sometime in the early 1940s. From 1939 through 1945, the Champion ran between Miami, Florida, and New York, New York, making stops along the way. (Courtesy of Buddy Hill.)

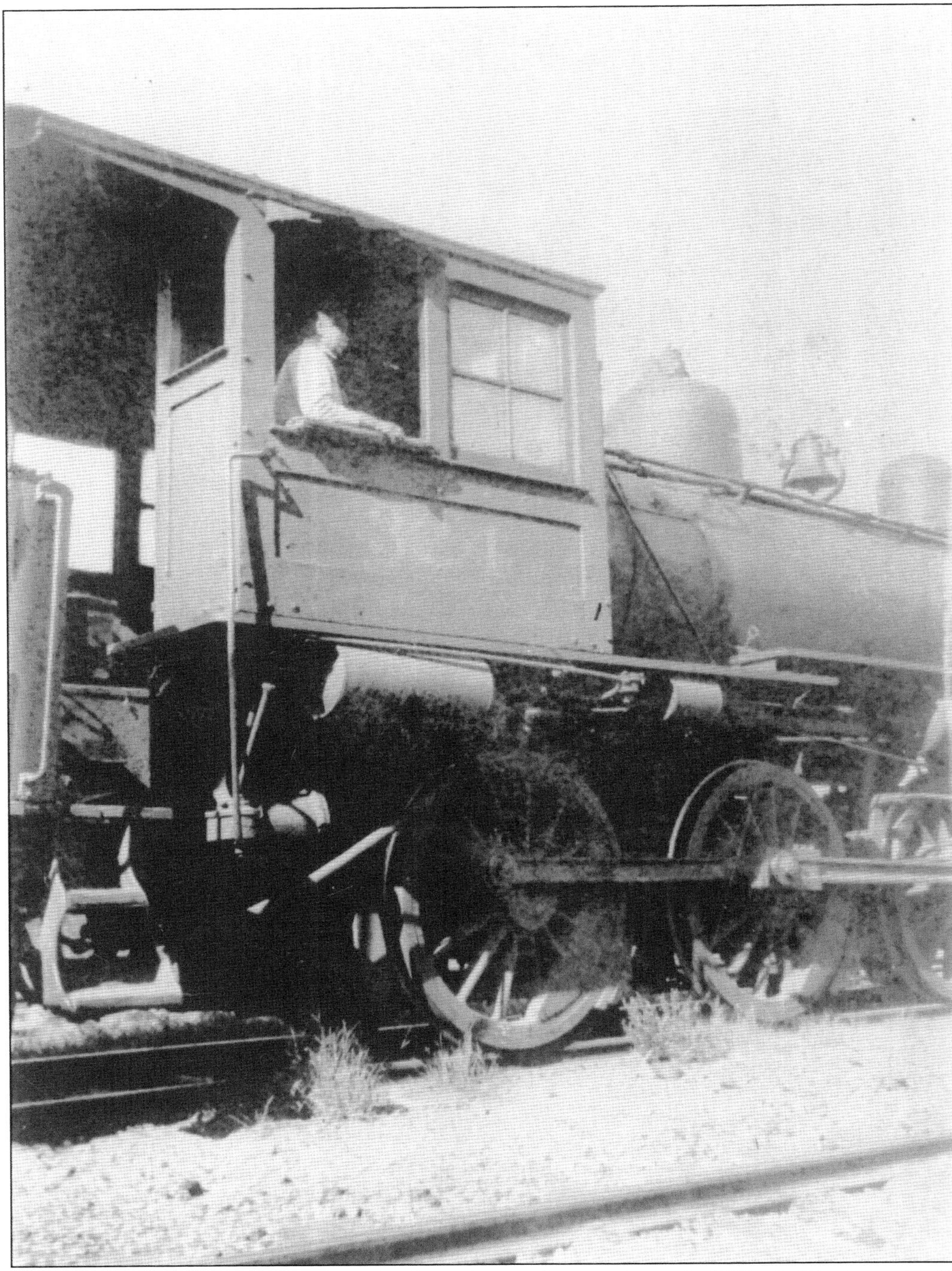

In this photograph, which is believed to have been taken in the early 1920s, is "Engineer Huff," who was more often than not found conducting many trains that ran through Ravenel. During this time, locomotive engineers typically specialized in 100 to 150 miles of track, so they knew the location of every signal, every turn, or the slightest change in the grade of the land, enabling them to control the train safely. (Courtesy of Buddy Hill.)

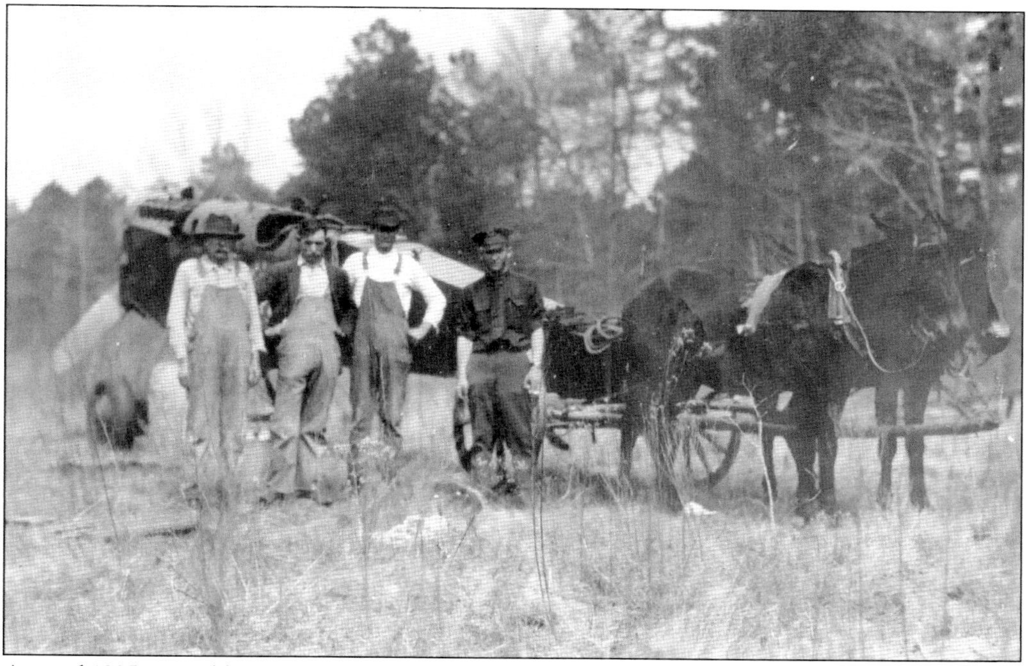

Around 1925, a World War I–era airplane on a training mission went down in the Eureka Township of Parker's Ferry in St. Paul's Parish. Because of rudimentary aviation engineering, these planes were much more fragile than planes today and oftentimes broke apart or encountered mechanical problems while in flight. The accident occurred between Robby Perry's house on Parker's Ferry Road and the Edisto River. In the photograph above taken by the military are, from left to right, George S. Garvin, Jessie Weeks, Robert "Robby" Perry, and an unidentified pilot. (Above, courtesy of Wilbur Jones; below, courtesy of Bobby Garvin.)

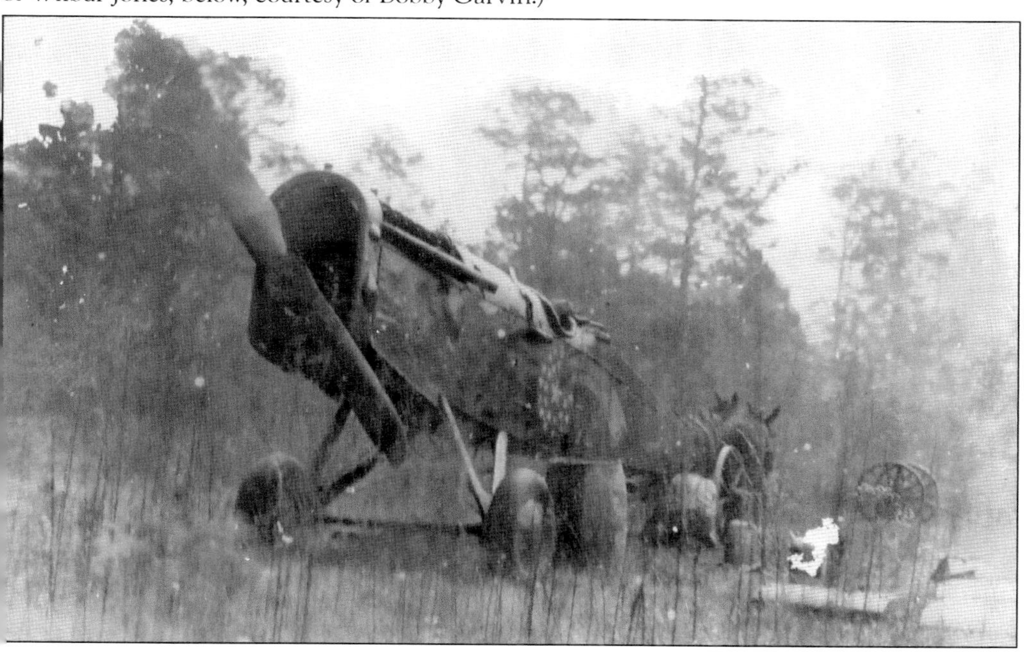

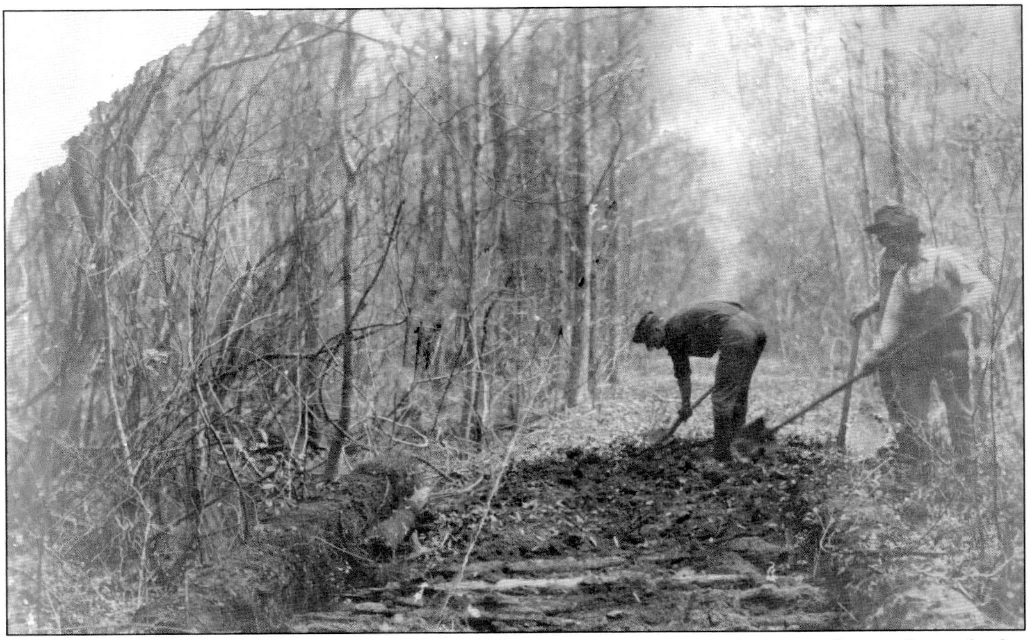

In this 1925 photograph are three unidentified men attempting to clear a pathway through the underbrush and marsh along the Edisto River to allow a team of horses to drag the plane out of the marsh where it had crashed and into town where it could be loaded onto the railway. (Courtesy of Bobby Garvin, Joe Weeks, and Johnny Garvin.)

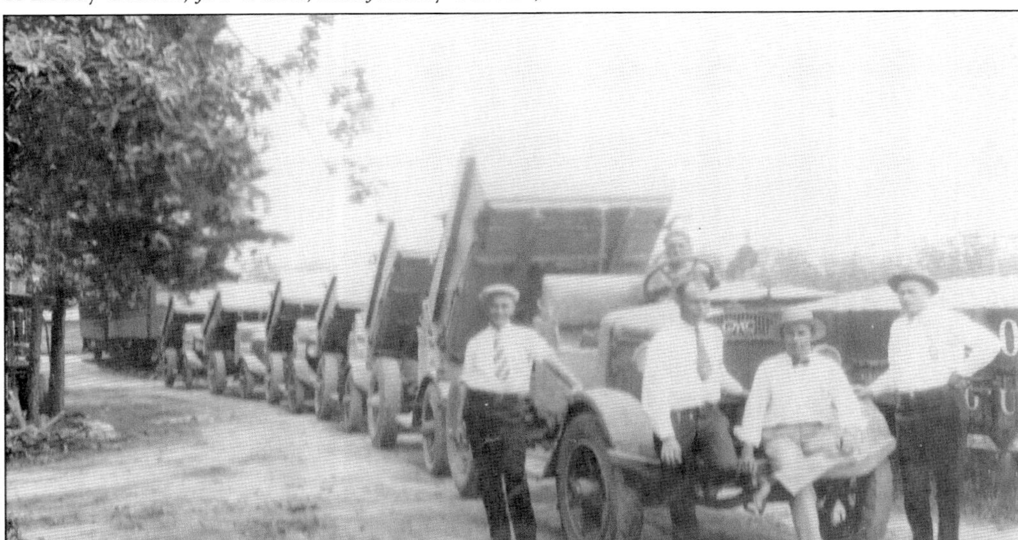

Here, Harvey H. Weeks (second from left), of Osborn, South Carolina, is shown with members of a construction crew and their General Motors trucks in the 1920s. While railroads had historically been the most-used form of transportation, the construction of roads became more necessary by the mid-1920s as cars increased in popularity. (Courtesy of Elizabeth "Betsy" Waters.)

Griner Chevrolet Company

Sales Service

Telephone 13

Hollywood South Carolina

Above is an advertisement for Griner Chevrolet from the St. Paul's High School yearbook. This dealership was located in Hollywood and was owned and operated by local businessman John Griner. Below is a picture of the inside of the business featuring a row of brand-new 1938 Chevys that had just arrived and likely created a buzz around town. (Above, courtesy of Mona Butler Williams; below, courtesy of David C. Reaves Sr.)

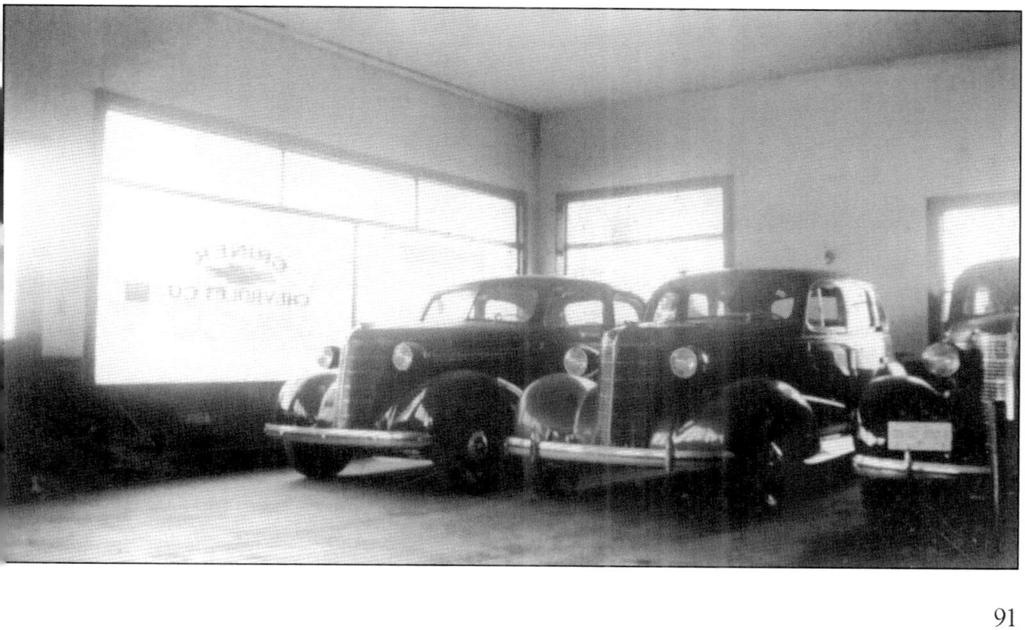

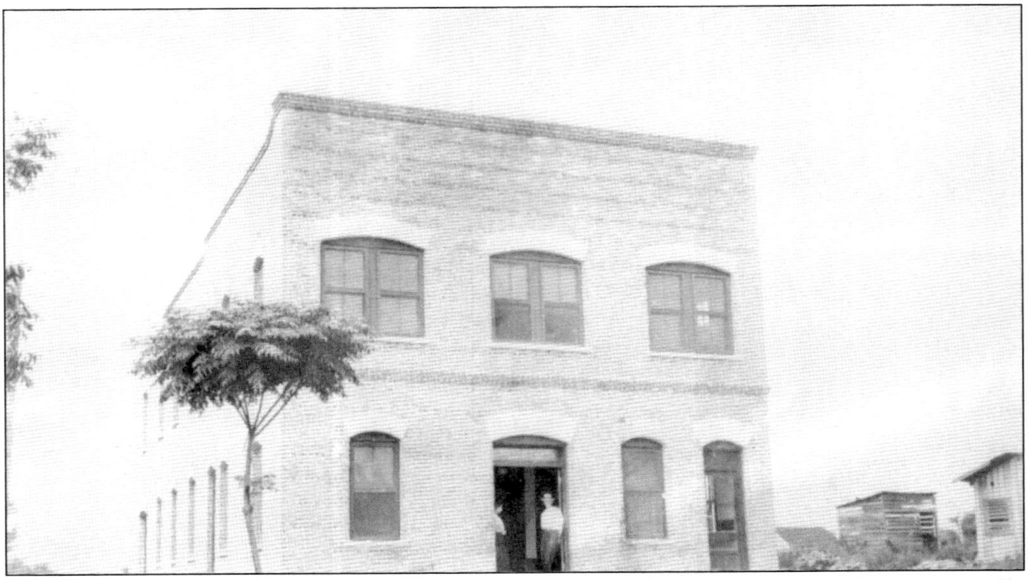

John Franklin Griner poses in the front of the Yonges Island Post Office in 1915. Griner originally came from Brunson, South Carolina, to work as the postmaster at Yonges Island. Later on, he started Griner Chevrolet in Hollywood. Griner was married to Marie Kulinsky and had two children—John Franklin Jr. and Antoinette (known as "Net"). (Courtesy of Anne Sanders Groover.)

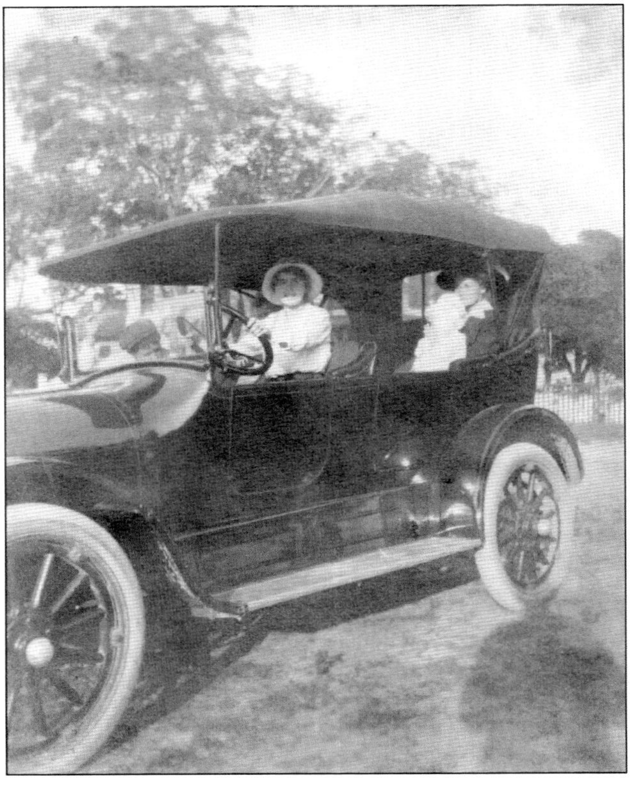

Antoinette "Net" Griner is driving her father's new car to town with unidentified passengers, thought to be family members. John Franklin Griner, who owned and operated Griner Chevrolet in Hollywood, always told his family, "I sell Chevys, but I love Buicks." The family has debated for years whether or not this is a Chevy or a Buick. The picture was taken in the mid-1920s. (Courtesy of Anne Sanders Groover.)

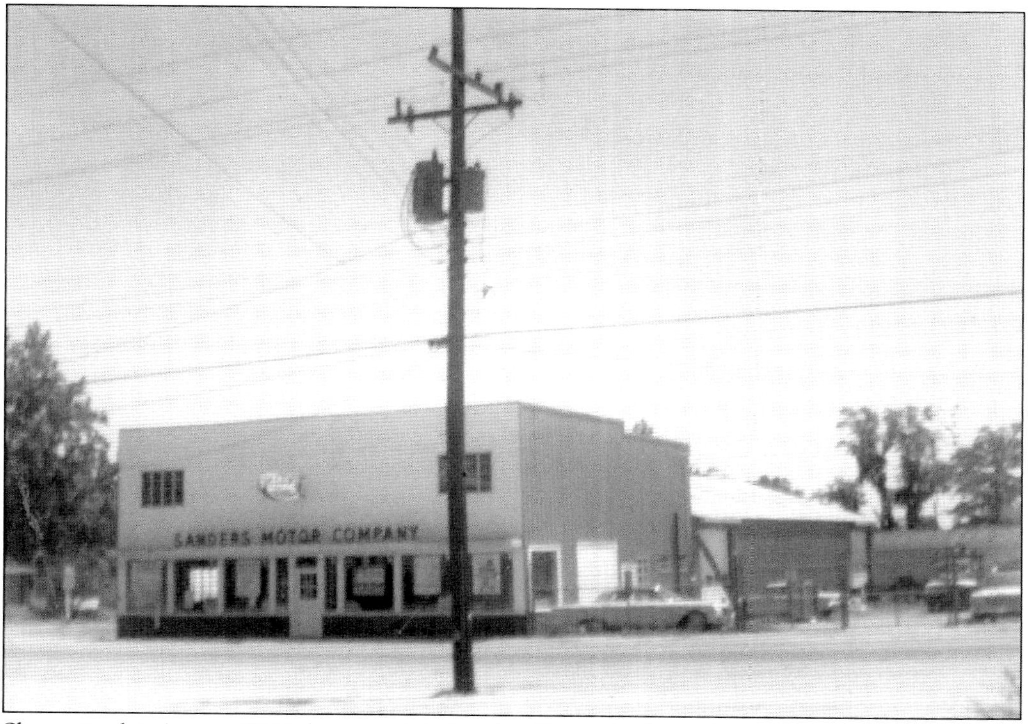

Shown in this August 1962 photograph is Sanders Ford, which was located at the present-day site of Robert Sarco Antiques. The Ford dealership was run by Sammy Sanders from the early 1940s until his untimely death in 1957. After Sammy's death, his wife, Net Griner Sanders, continued to run the business and was the only woman who owned a car dealership in South Carolina at the time. Net eventually sold the dealership to Norman Sires, who had managed the business for her after Sammy's death. (Courtesy of Anne Sanders Groover.)

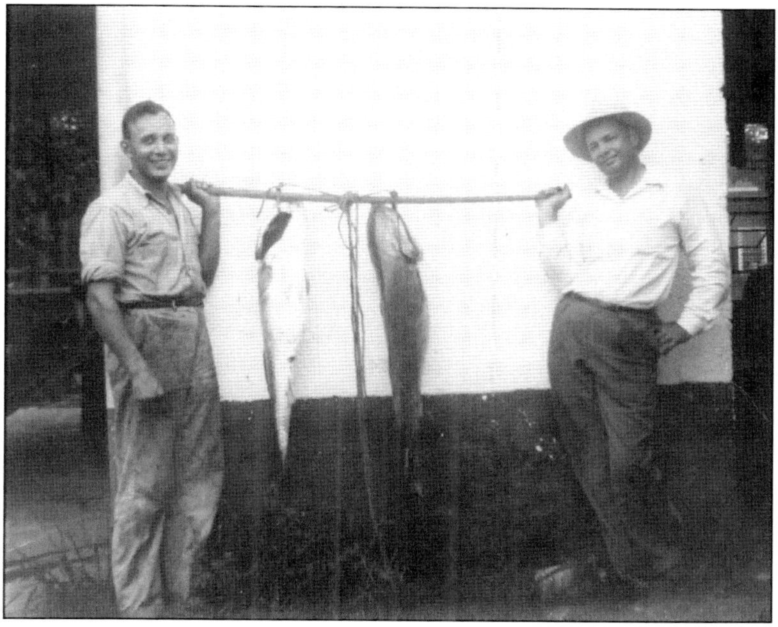

Neal Golden (left) and Sammy Sanders are showing off their catch after a fishing trip down toward Botany Bay one morning in the early to mid-1950s. They are standing in front of Neals Body Shop, which was adjacent to Sanders Ford in Hollywood, South Carolina. (Courtesy of Anne Sanders Groover.)

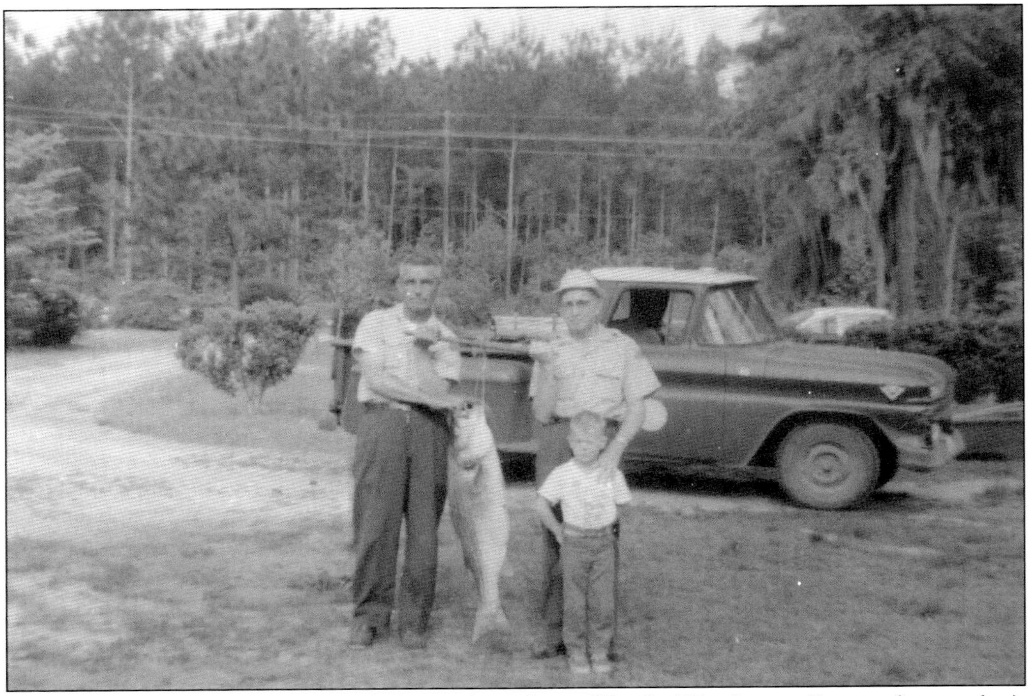

In this 1966 photograph are, from left to right, Johnny Woods (Charleston County fire warden), John Irvin Garvin (Charleston County fire ranger), and his grandson Robby Garvin. They are standing in front of a South Carolina State Fire Ranger pickup truck parked at the Adams Run fire tower in Osborn after a day of fishing. (Courtesy of Bobby Garvin.)

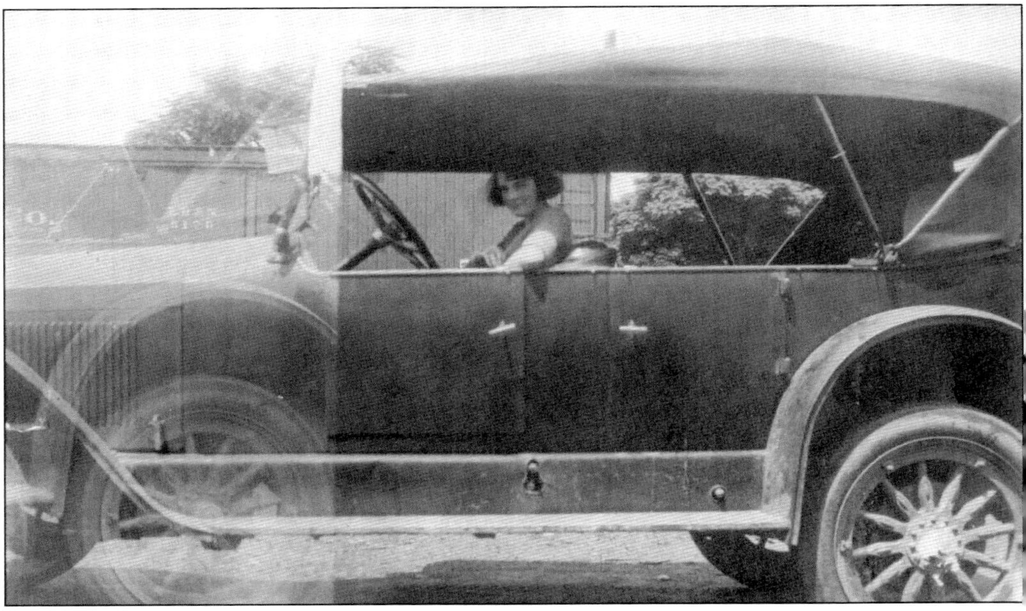

In this 1902 photograph, Weston Harrison Geraty, known to everyone in the community as "Duck," is caught by an unidentified photographer getting ready to drive her first convertible around town. (Courtesy of William "Billy" Geraty.)

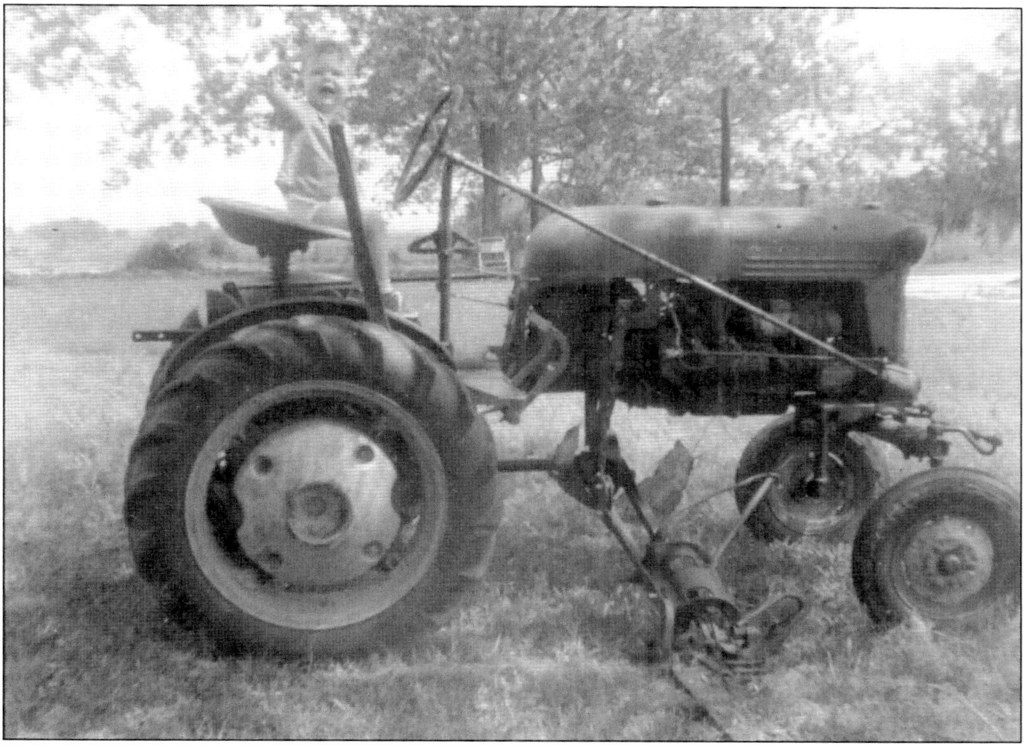

Pictured in 1956, Larry Commins (age two) is sitting on one of the earliest tractors in Meggett. It was used to cut grass both in the Commins' yard and around town. Typically, it would be parked in the front yard of the Commins house off Highway 165. (Courtesy of Anne Commins.)

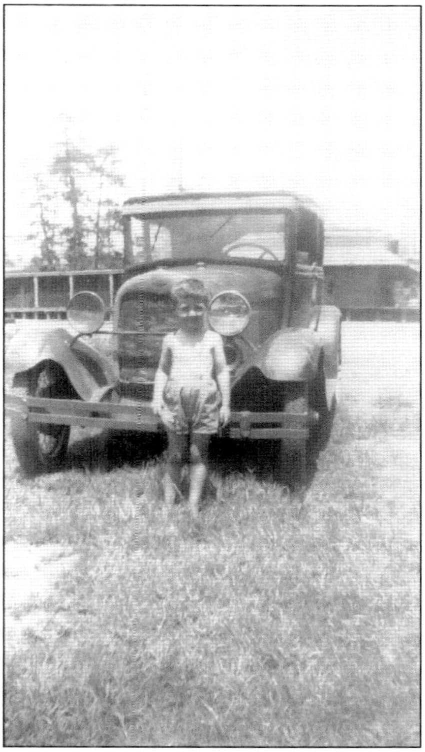

David C. Reaves Jr. is standing in front of a Model A Ford in 1945. The car is parked behind the Hollywood Fountain drugstore and in front of the Hollywood Icehouse. Anne Reaves Commins, who would have been 14 at the time, can remember driving this car and her younger brother home from the drugstore (and singing "Give Me Five More Minutes") if her parents were working late. (Courtesy of Anne Reaves Commins.)

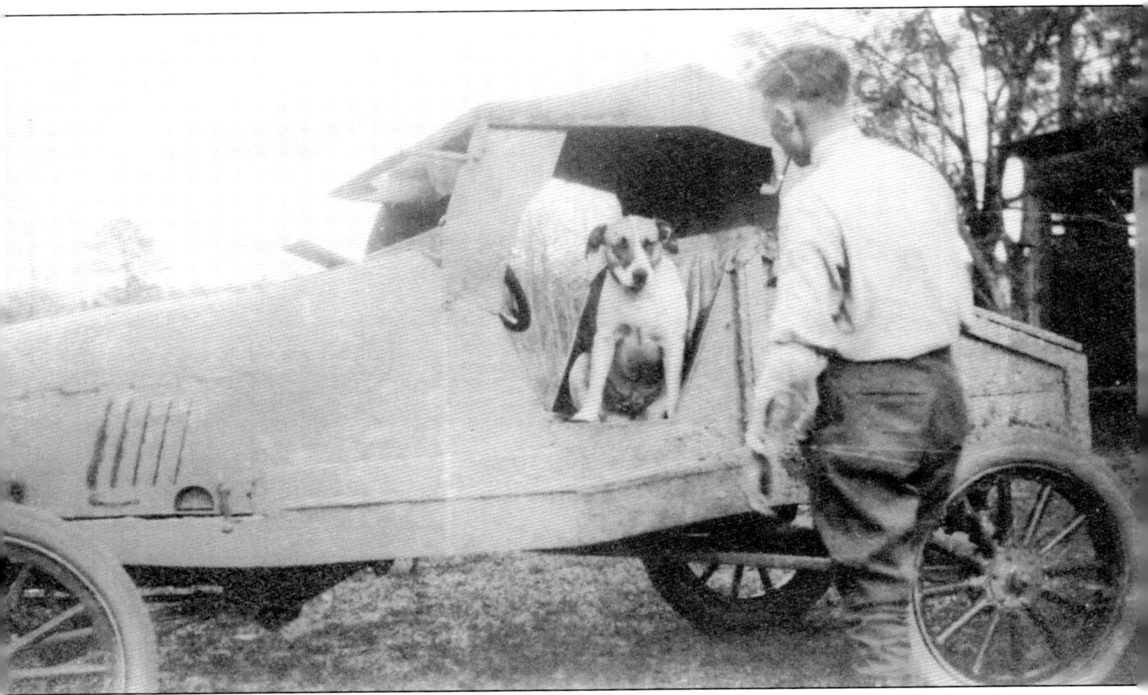

Milton "Carroll" Boineau (1895–1963) is pictured in 1925 with his beloved American bulldog, Rex, who died in 1926, shortly after this picture was taken. Carroll and Rex were often seen driving around town in this "improvised" convertible. Carroll began as a mail carrier in Meggett but went on to become the mail carrier out of the Osborn Post Office for 44 years. (Courtesy of Jackie Boineau.)

In the photograph at right is William "Bill" Barnwell, brother of Osborn "Ozzie" Barnwell. Bill was called to duty and reported to Fort Bragg, North Carolina, in January 1941. He served with the 813th Tank Destroyer Battalion, 2nd Army Corps and participated in the 1942–1943 North Africa campaign during World War II. In addition, he landed in France on June 6, 1944. Bill retired from the Army as a lieutenant colonel. Below is a gas ration coupon belonging to Joe F. Towles in 1943. During World War II, many consumer items, such as gasoline, were rationed across the United States as the country fought the war. (Right, courtesy of Rosanne King Jones; below, courtesy of Beth Van Nordt.)

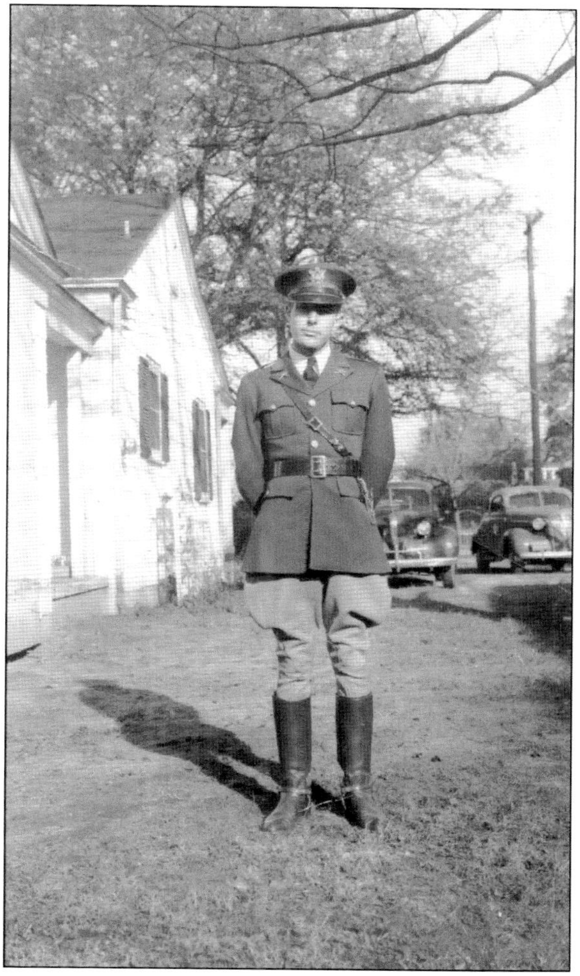

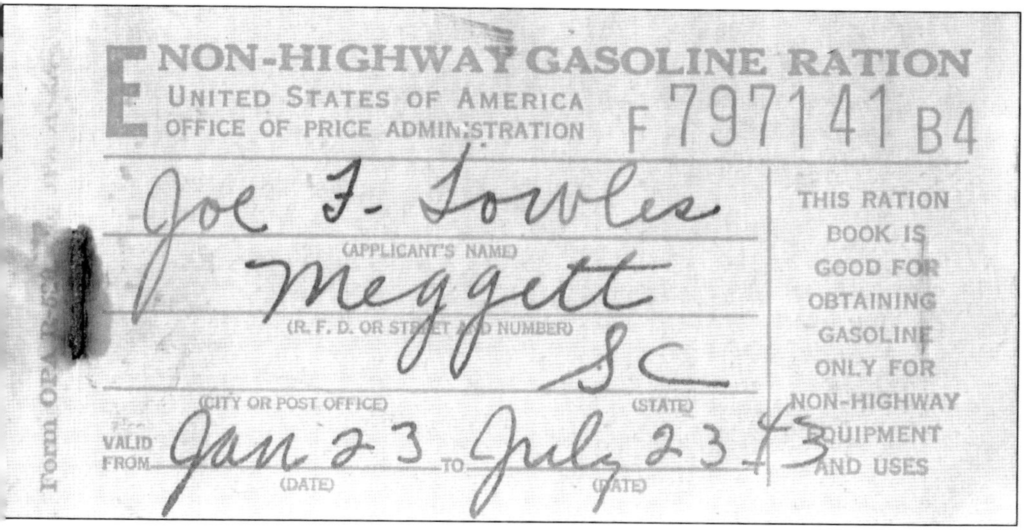

E NON-HIGHWAY GASOLINE RATION
UNITED STATES OF AMERICA
OFFICE OF PRICE ADMINISTRATION
F 797141 B4

Joe F. Towles
(APPLICANT'S NAME)
Meggett
(R. F. D. OR STREET AND NUMBER)
SC
(CITY OR POST OFFICE) (STATE)
VALID FROM Jan 23 (DATE) TO July 23 43 (DATE)

THIS RATION BOOK IS GOOD FOR OBTAINING GASOLINE ONLY FOR NON-HIGHWAY EQUIPMENT AND USES

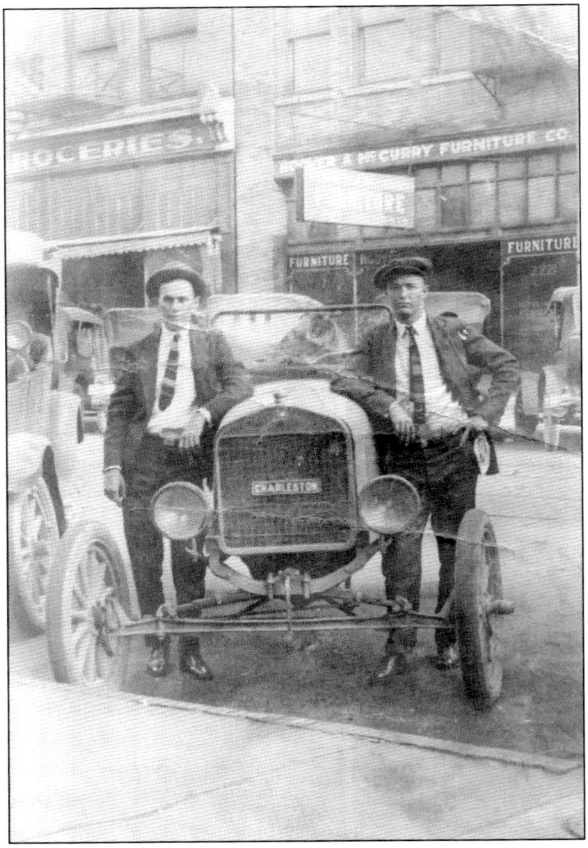

At left are St. Paul's Parish boys Newton Butler (left), of Ravenel, and Holland Baldwin, who drove their car all the way to Texas in the mid-1920s. A Charleston plate is attached to the front of the car. Below is Butler's registration for his "Ford Motorcar," dated July 21, 1916. At that time, the cost to obtain a license plate was determined by the horsepower of a vehicle and not the weight as it is today. Residents were required to pay 25¢ per horsepower. The plates produced at that time had the expiration date as well as the county and the state stamped on them. (Both, courtesy of Mona Butler Williams.)

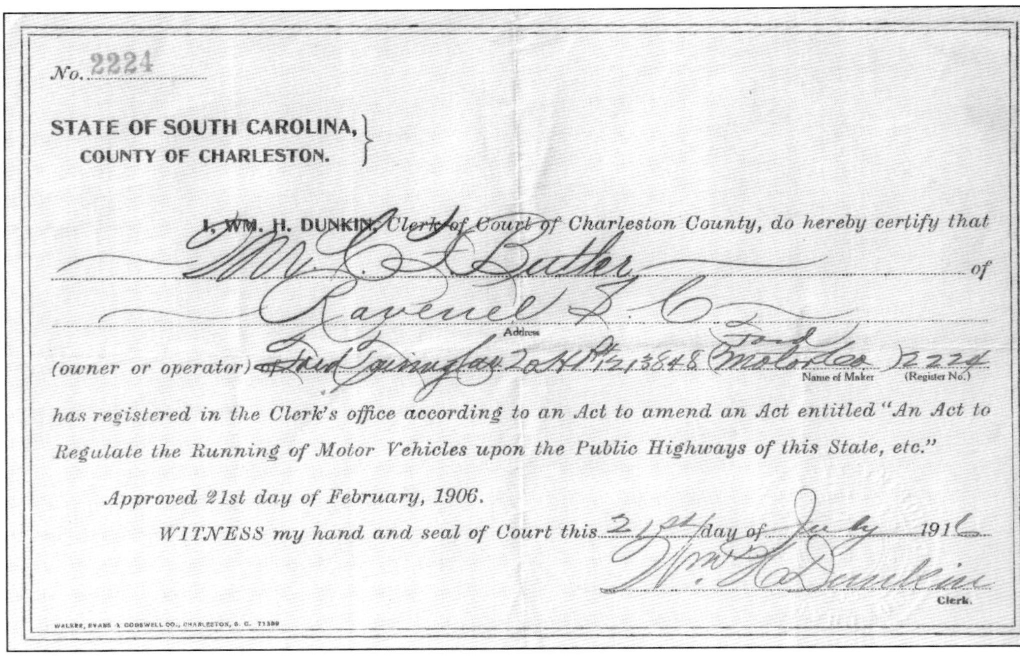

No. 2224

STATE OF SOUTH CAROLINA,
COUNTY OF CHARLESTON.

I, WM. H. DUNKIN, *Clerk of Court of Charleston County, do hereby certify that* Mr. C. L. Butler *of* Ravenel S. C.

Address

(owner or operator) Ford Touring car 20 H P No. 213848 Ford Motor Co 2224

Name of Maker (Register No.)

has registered in the Clerk's office according to an Act to amend an Act entitled "An Act to Regulate the Running of Motor Vehicles upon the Public Highways of this State, etc."

Approved 21st day of February, 1906.

WITNESS my hand and seal of Court this 21st *day of* July *191*6

Wm. H. Dunkin
Clerk.

WALKER, EVANS & COGSWELL CO., CHARLESTON, S. C. 71389

Four

Parish People

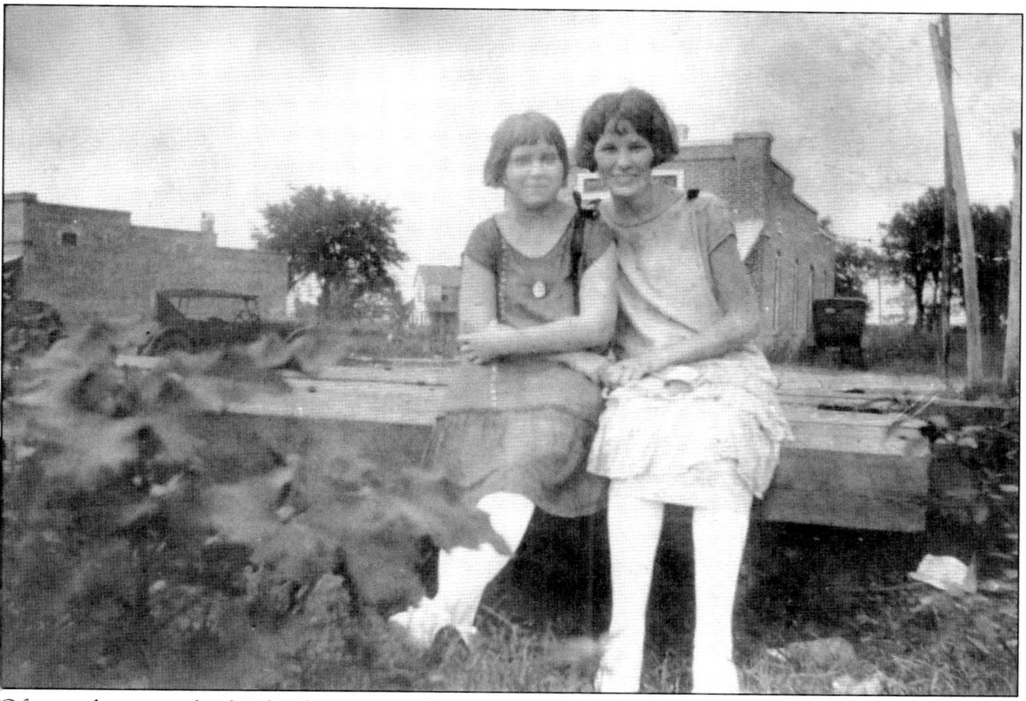

Often, when people think of a town, village, or community, individual faces or names come to mind. A town is only as good as the people who live within it. St. Paul's Parish contains many small towns and communities, all of which are home to numerous families and individuals. Sitting across from the Exchange Bank/Southern Produce Association Building around 1926, Mary Dodd (right) and an unidentified friend pose for the camera. Dodd was a senior at Meggett High School, played on the basketball team, and was thought to be very musically talented. (Courtesy of Anne Commins.)

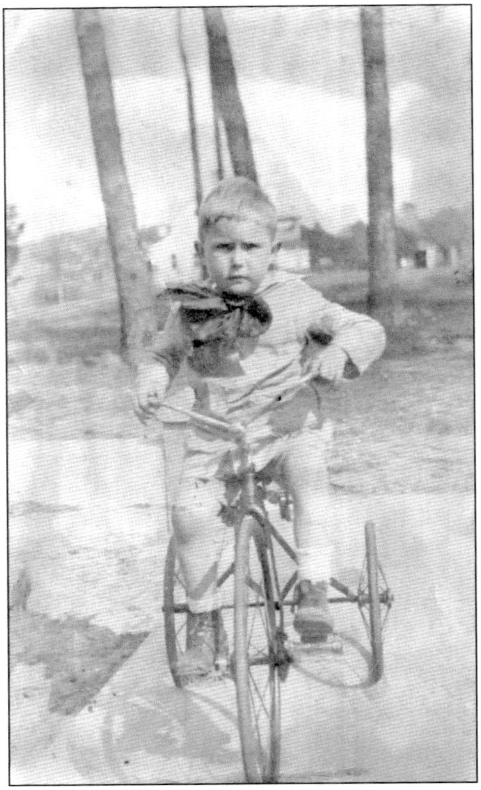

Two-year-old Josiah Francis Towles (1913–1961) is riding his tricycle in front of Quigley House around 1915. Josiah was the son of Daniel Quigley Towles Sr. and Mary Enid Mixon. Josiah would go on to marry Alma Rogers Gilmore and have five children. (Courtesy of Beth Van Norte.)

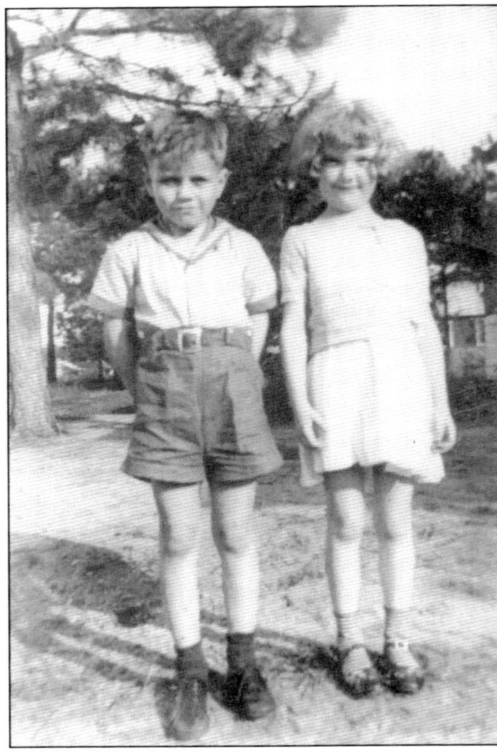

Carl Butler Jr. and Anne Reaves Commins (both age six) are photographed in 1936 by Anne's father after walking to meet each other. The two parish children were good friends and playmates. (Courtesy of Anne Commins.)

James Davis, cousin of David C. Reaves Sr., stands outside of the Meggett Drug store and one of the Meggett grocery stores in 1926–1927. Davis was down visiting from Sumter, South Carolina. Behind Davis are other businesses, including another grocery store and a drugstore and feed store owned by W.C. Geraty. (Courtesy of Anne Commins.)

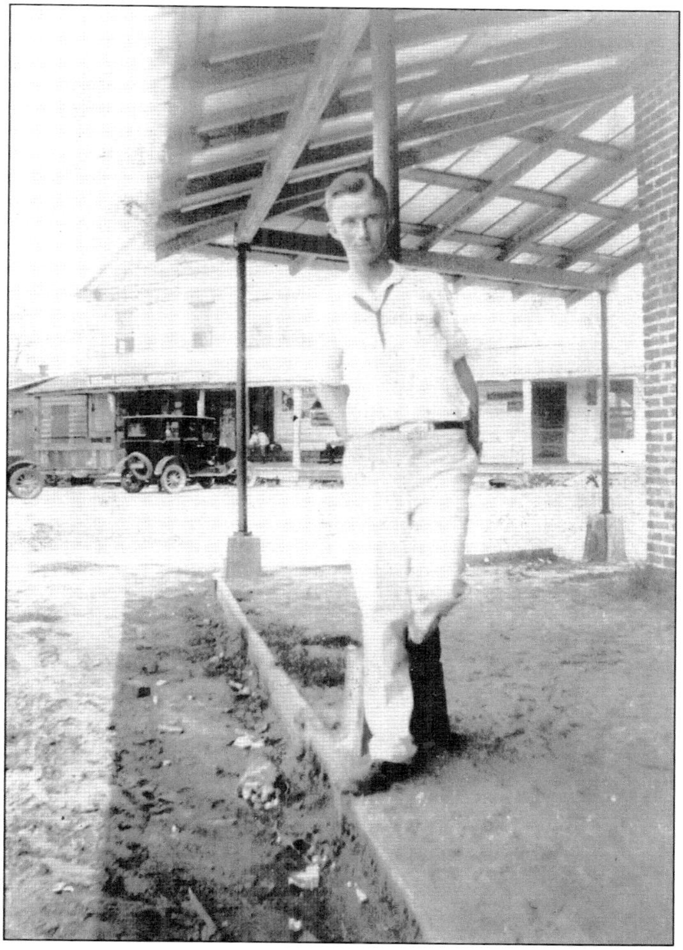

In the picture below is Bessie King with a group of "King kids." From left to right are (first row) Rosanne and Connie; (second row) Harriet, Mary Virginia "Ginny," Lawrence, and Betty; (third row) Bessie. The girls had gathered for a friendly game of baseball in 1948, and as memory recalls, Betty was typically the one in charge of the crew. (Courtesy of Rosanne King Jones.)

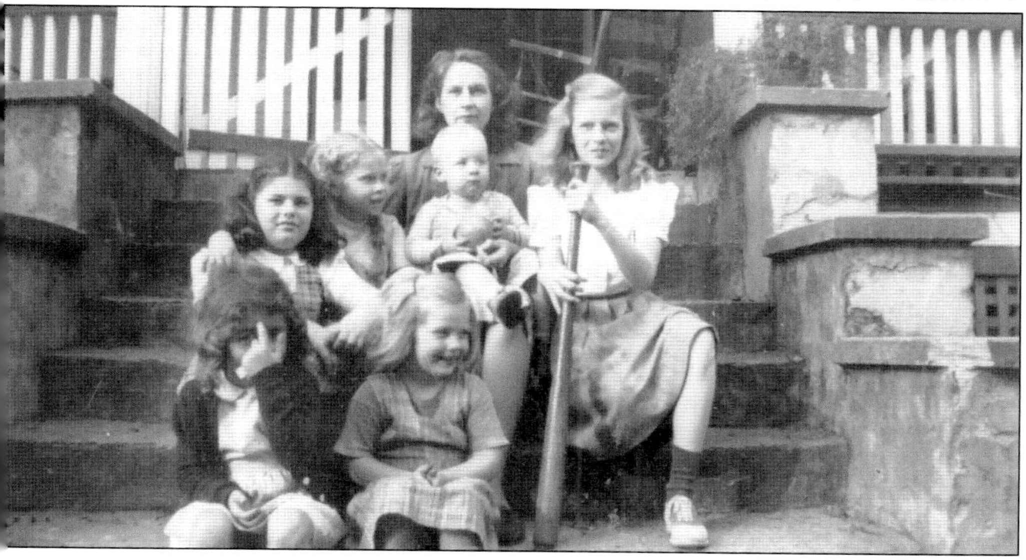

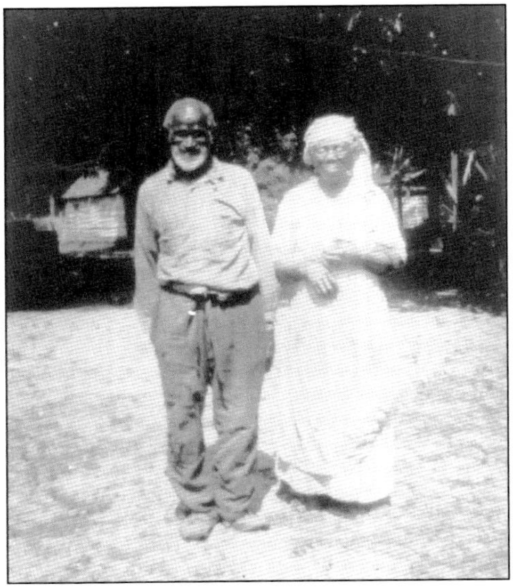

In this early-20th-century photograph are Ishmael Washington and Philippa Fields, known as "Maum Philippa." Maum Philippa worked for Anna Cuthbert Taylor, Dr. Taylor's wife, in Adams Run. Philippa's husband, Joe Fields, was the personal servant to South Carolina governor Duncan Clinch Heyward (1903–1907). Both Philippa and Joe Fields are mentioned in *The Black Border: Gullah Stories of the Carolina Coast* by Ambrose E. Gonzales. (Courtesy of Anna Blythe.)

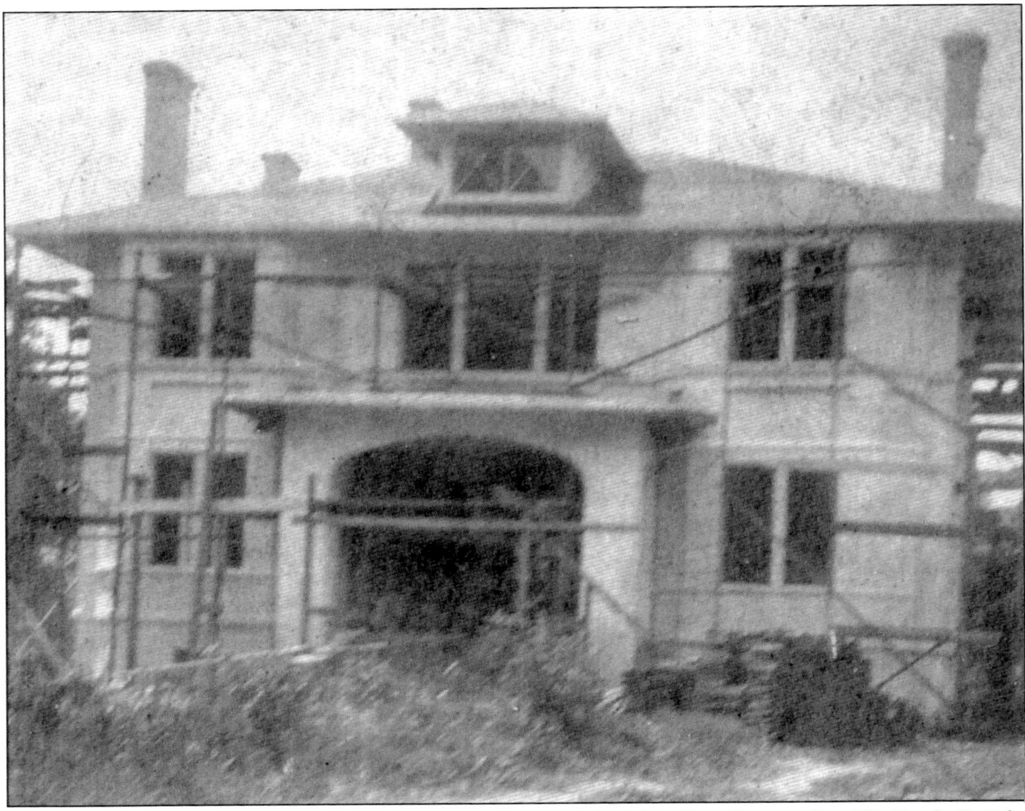

Oaklawn Plantation was otherwise known as the "Gonzales Place" to the people of St. Paul's Parish. Oaklawn is located near the Adams Run/Parker's Ferry area of St. Paul's Parish and was home to the Gonzales family. Perhaps the most famous of the Gonzales clan is Ambrose Elliott Gonzales (May 27, 1957–July 11, 1926), who with his brother started the *State* newspaper in Columbia, South Carolina, and was also a highly acclaimed published author. (Courtesy of Bobby Garvin.)

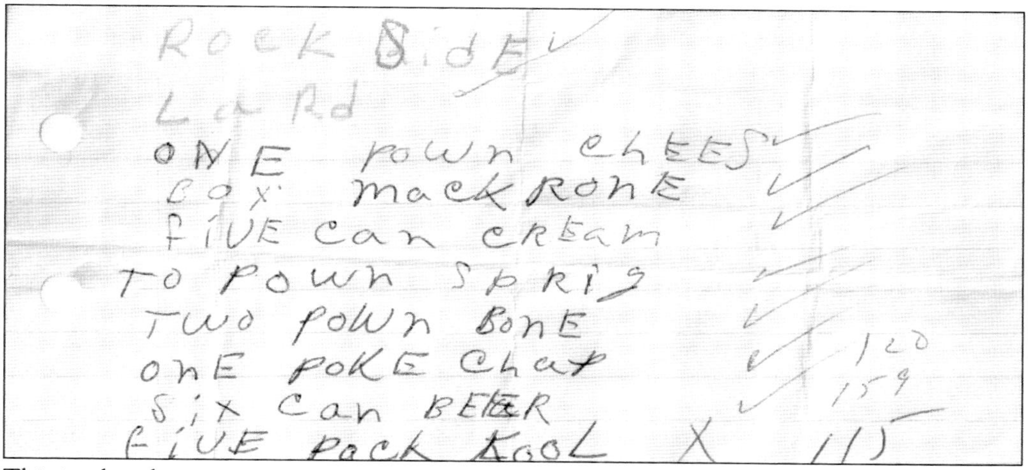

ROcK SidE
LaRd
ONE Pown cHEES
BOX mackRONE
fIVE can cREam
TO Pown SpRig
TWO pOWn BONE
ONE POKE Chat
SIX Can BEER
FIVE Pack KOOL

This is a handwritten note in Gullah that was given to Ozzie Barnwell at the Roxbury Mercantile by a young African American child. The child had been sent to the store to collect some goods that the family needed. (Courtesy of Joyce "Dumpy" Towles Barnwell Cave.)

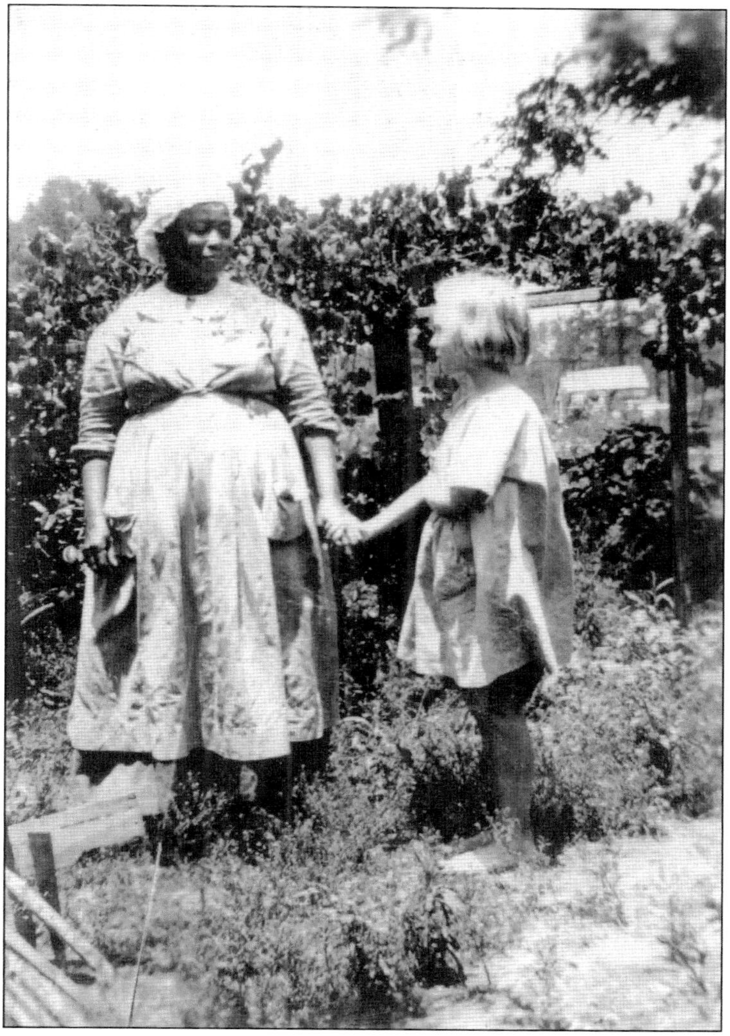

Seen here in the early 1920s are Dianna Fraizer (left) and Dorothy Kershaw, first cousin to Jackie Boineau. Fraizer was considered a very loyal family "helper." She looked after the family's children and worked around the house during the daytime. Dianna had a house in Adams Run and walked to work every day. (Courtesy of Jackie Boineau.)

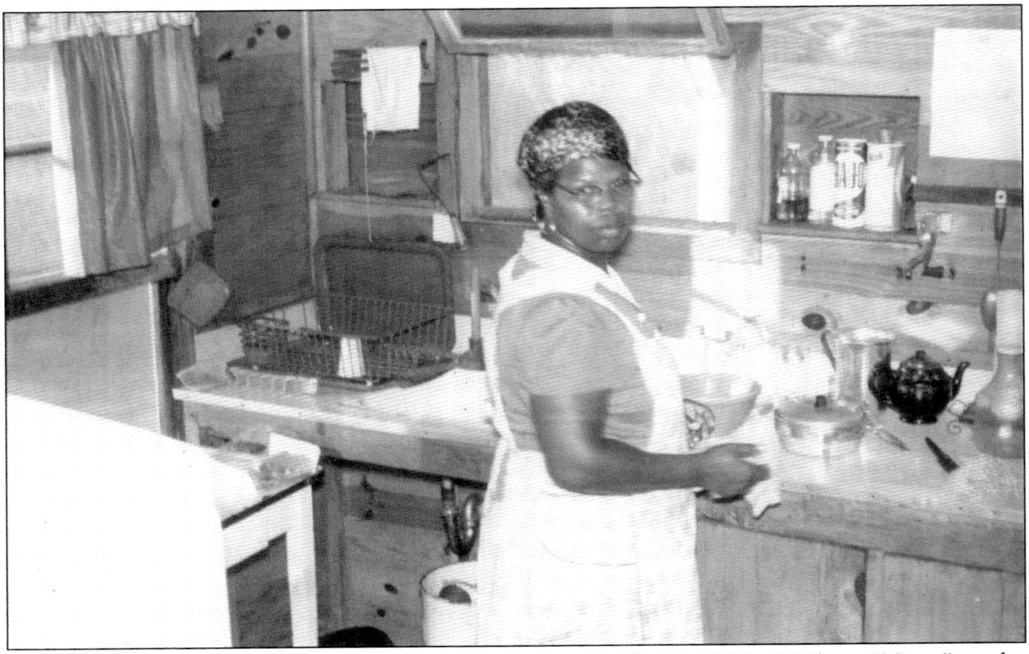

Seen at an Edisto Beach house in 1961 is Viola Smalls Fludd (1906–1980), or "Nina," as she was known to generations of Butler children. Fludd was a lifetime resident of St. Paul's Parish. (Courtesy of Beth Van Norte.)

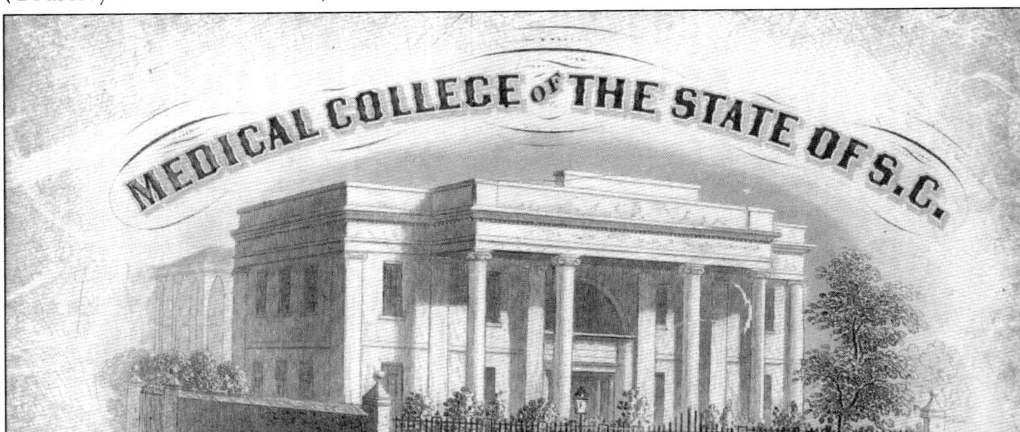

Considered the first medical school in the South, the Medical College of South Carolina opened its doors in 1824. Its graduated its first class of five students in 1825. The school opened its first permanent building in 1826 at Queen and Franklin Streets. Today, it is known as the Medical University of South Carolina (MUSC). This is the matriculation card for Dr. Taylor from Adams Run. (Courtesy of Anna Blythe.)

Dr. Joseph "Tally" Taliafero Jr. was a physician in Adam's Run. Taliafero attended Porter Military Academy (later Porter-Gaud), Hobart College in New York State, and then completed medical school at the Medical College of South Carolina before he purchased Dr. Grimballs's practice in St. Paul's Parish. (Courtesy of Anna Blythe.)

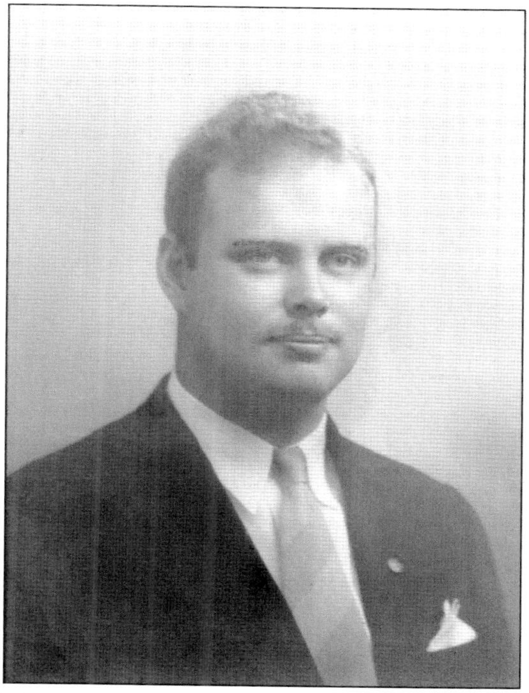

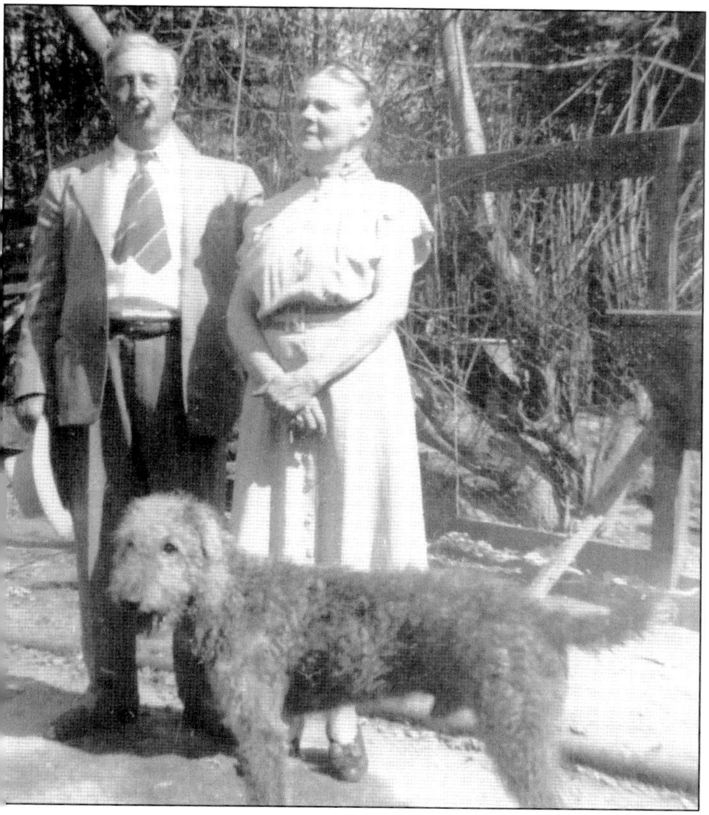

Dr. Richard Robert Prentiss and his wife, Claudia Wilkinson Prentiss, pose in the early 1930s along with Butch, one of the doctor's many dogs. Dr. Prentiss graduated from the Medical College of South Carolina in 1925 and opened up his own clapboard shop in his backyard soon after. Dr. Prentiss had many dogs, but all can recall Butch, as he is fondly remembered for eating the first pair of $50 shoes that the doctor ever owned. (Courtesy of Marsha Blandenburg.)

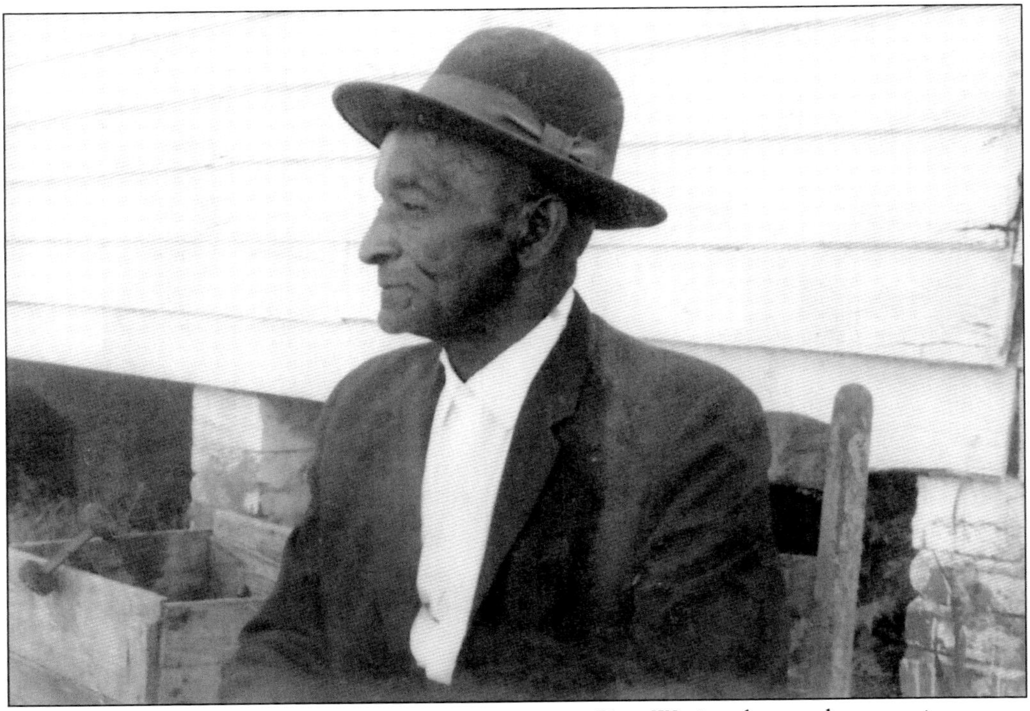

Tom Waring, known by many in the Adams Run community as "Ole Tom," is sitting beside his house in the early 1950s. Waring was highly respected by his peers. He did odd jobs, such as cutting and delivering wood, farming, and other handyman services, for families around the village. (Courtesy of Jackie Boineau.)

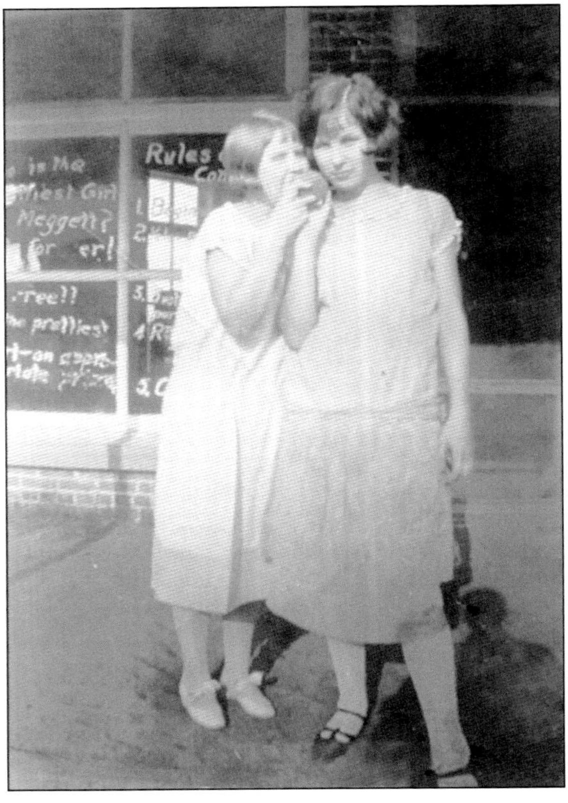

In this photograph taken in 1926–1927, Lavinia Hughes (left) and Mary Dodd are standing in front of the drugstore in Meggett. Behind the girls on the plate glass is writing announcing the Miss Meggett contest. (Courtesy of Anne Commins.)

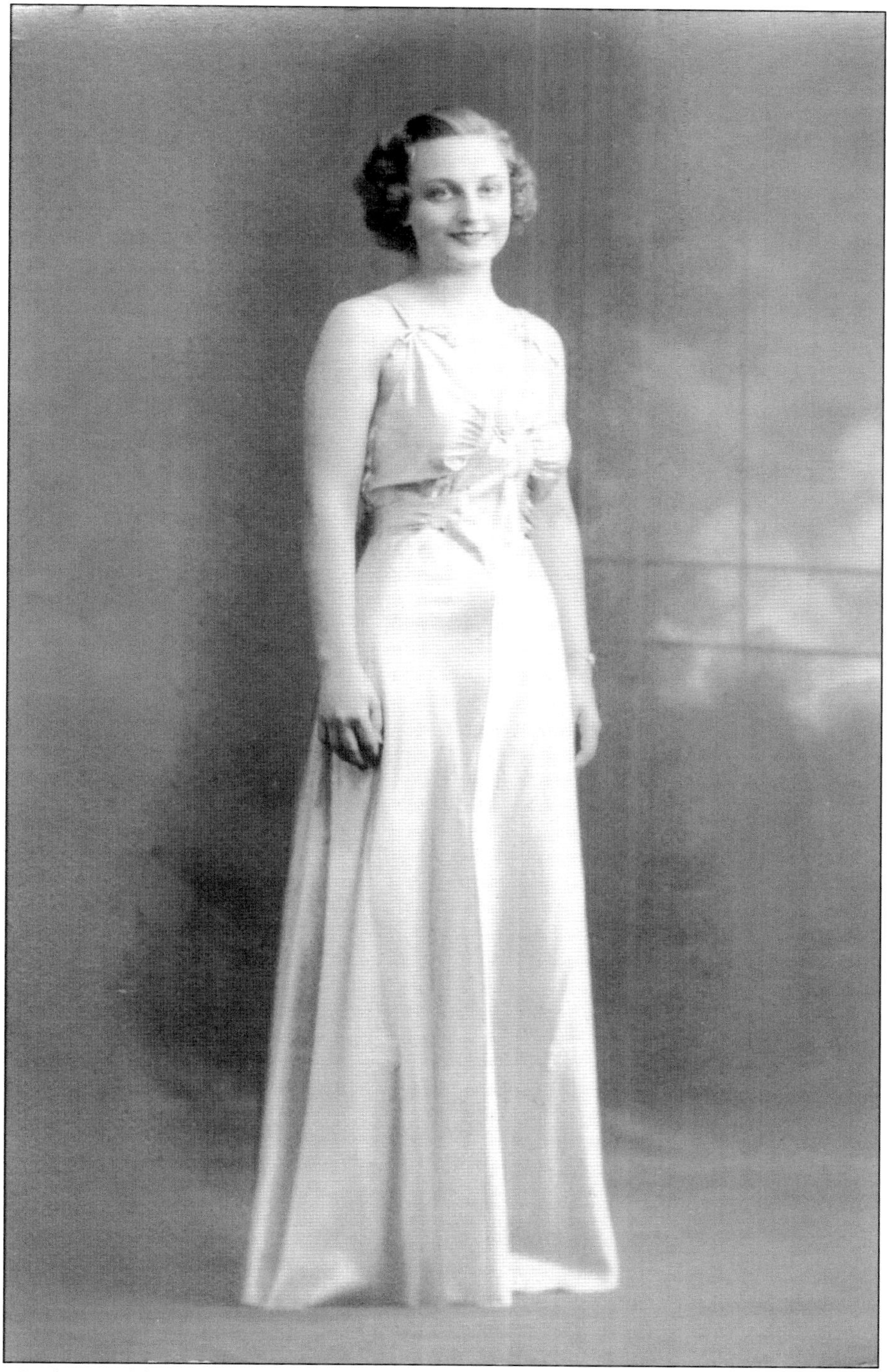

Naomi Bailey Welch is pictured after she has just been named Miss Meggett in the late 1930s. The pageant was held in Charleston and was an affiliate of the Miss America Pageant. During her reign, Welch represented Meggett in various capacities, such as a maid at the Sons of Confederate Veterans Reunion in Columbia, South Carolina. (Courtesy of David and Lisa Kennedy.)

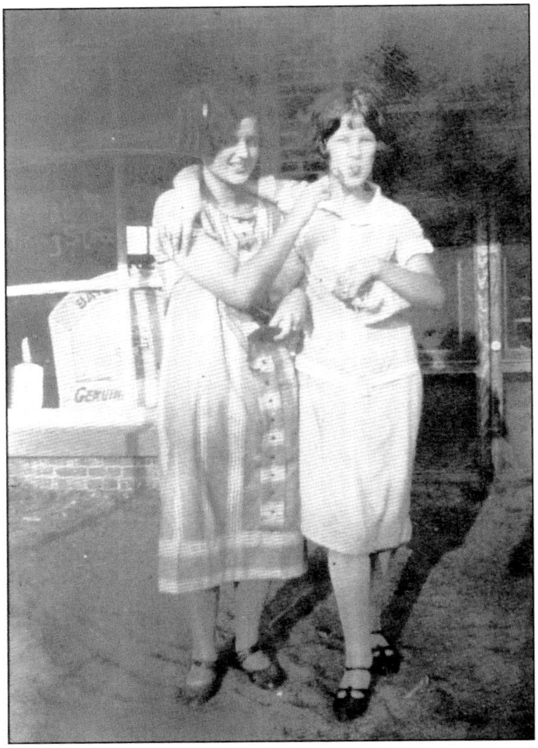

In a photograph taken in 1926–1927, Meggett schoolgirls and best friends Marion Stroble (left) and Mary Dodd are making faces at the camera. They are standing in front of a string of stores in downtown Meggett. In the photograph below, taken around 1926–1927, Enid Carr is standing just across the street from the same series of stores. She is holding a sign indicating that Charleston is 21 miles down the road. (Both, courtesy of Anne Commins.)

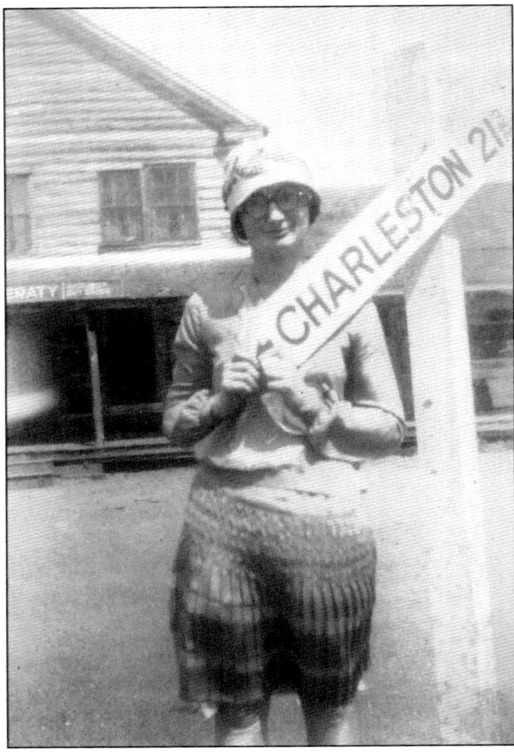

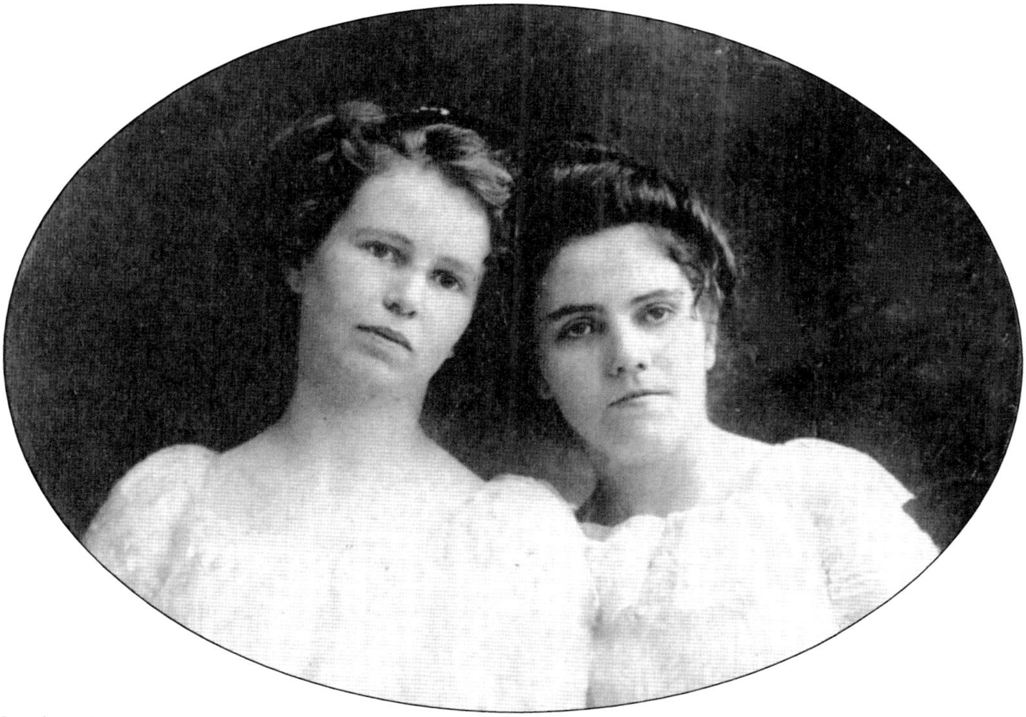

In the photograph above are Emily Bryan Jenkins (left) and Agnes Wilkinson Wilson, and in the photograph below are Claude LaRoche (left) and Mattie Bryan. The Bryan women were part of the larger Bryan family that owned and operated the Bryan Carbonating Works on Yonges Island off Toogoodoo Road. Both pictures are from around 1895–1900. (Both, courtesy of Marsha Blandenburg.)

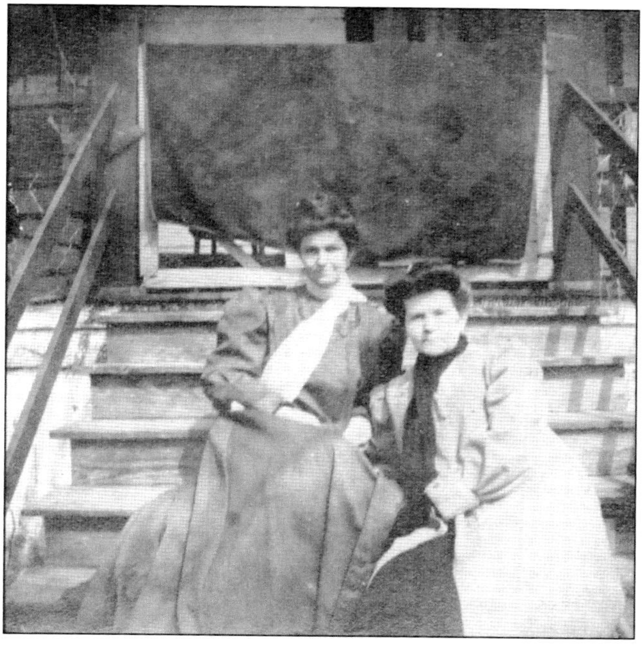

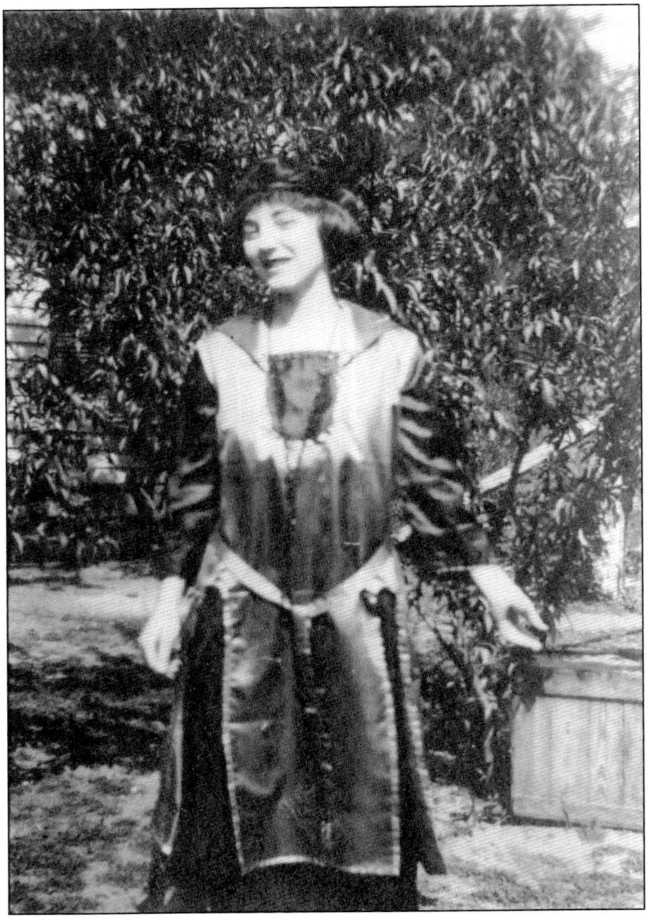

Kathleen Finch was the daughter of Reverend Finch in Adams Run. Reverend Finch was also the principal of Adams Run School. Kathleen is remembered as the young woman who was friends with everyone and called "Everybody's Girlfriend" by many. The photograph was taken in the mid-1920s. (Courtesy of Jackie Boineau.)

From left to right, Richard Tumblestone, Bub Tumblestone, and Milton Muckenfuss are pictured in the early 1920s. The boys are at Prentiss House on Toogoodoo Road in Meggett, attempting to enter a photograph in a contest held by Eastman Kodak. (Courtesy of Marsha Blandenburg.)

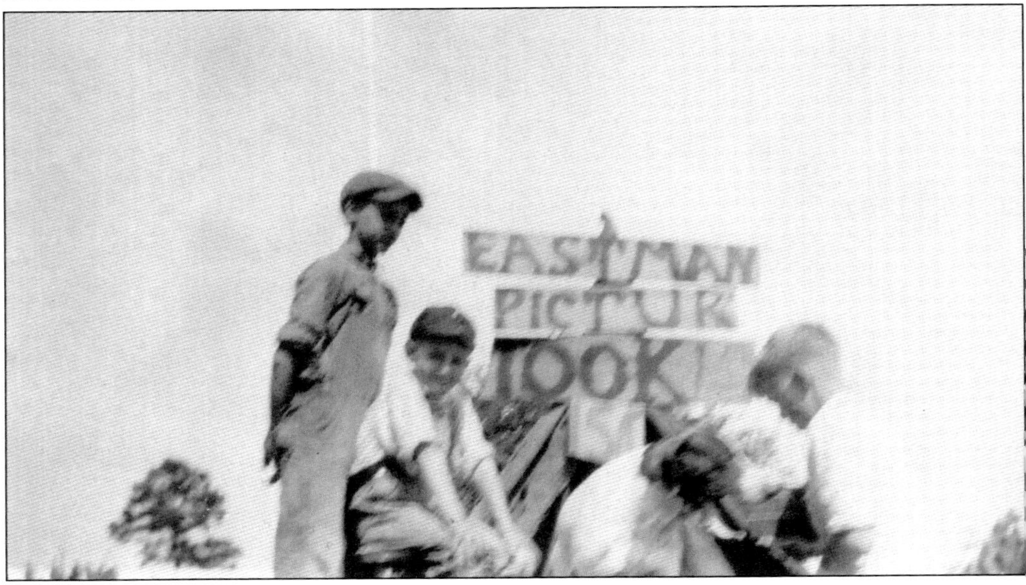

Butler girls Esther (left) and Francis pose for a picture while swimming. The girls lived in Ravenel with their two other sisters and mother and father. The date and location of photograph are unknown. (Courtesy of Mona Butler Williams.)

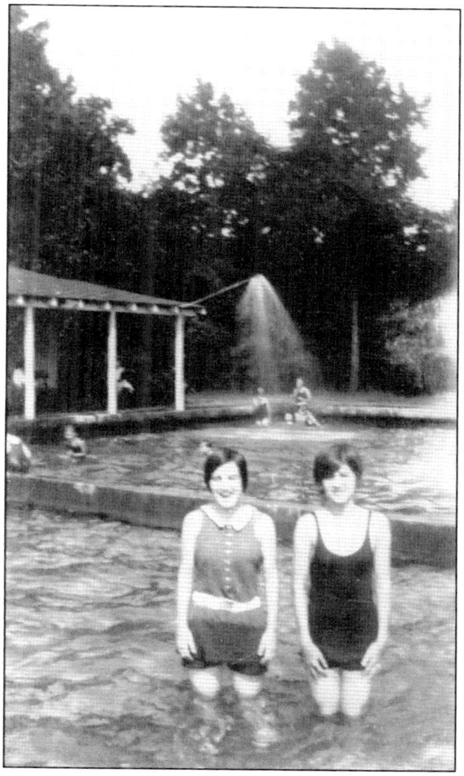

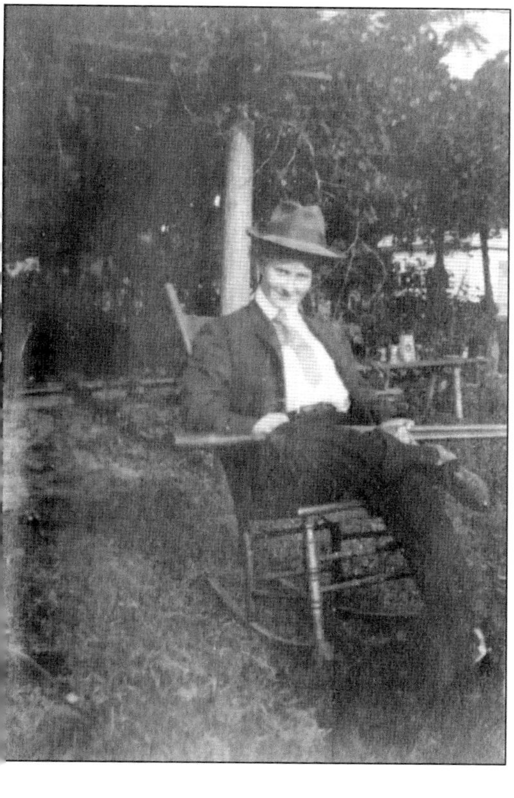

In this picture, labeled "A Visitor," Hilmer "Hilly" Behling is dressed up in Dowey Behling's clothes in the summer of 1909. Her relatives recall that Hilly loved to dance and was fun-loving. This image is an example of her jovial spirit, as it was not common to find very many women dressed up in men's clothing in 1909. (Courtesy of Sarah Roberts.)

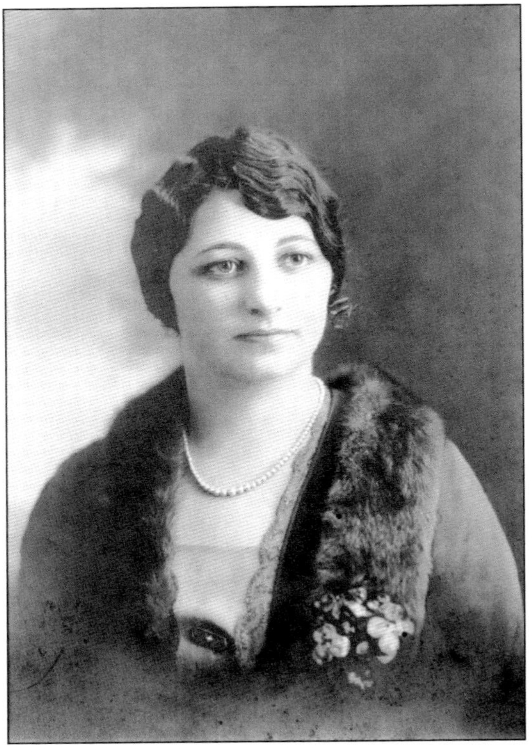

In the portrait at left is Lena Butler Joyner (1900–1993), sister of Esther and Francis. She was the wife of Rev. Ernest Luther Joyner. Below is Emilie Butler Towles, sister to Lena, Esther, and Francis. Emilie was first married to Ferdinand Winfield Towles, and after he died, she married his cousin John Oliver Towles. (Both, courtesy of Mona Butler Williams.)

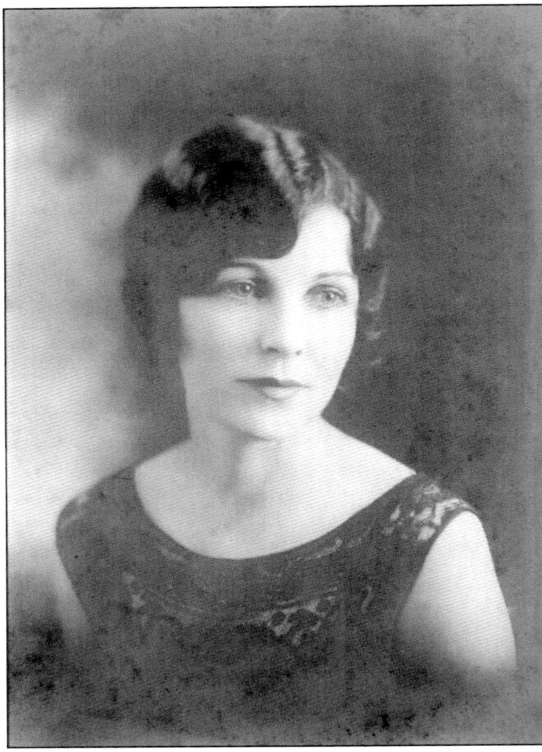

Seen at right is a young Ludar Cadow Behling Jr., affectionately called "Bubba" or "Little Dowie" by his relatives and friends. A caption on the 1912 photograph states that he is "on the lam." Dowie was "running away from home," but had remembered to take off his "good" shoes when he was crossing the marsh. (Courtesy of Sarah Roberts.)

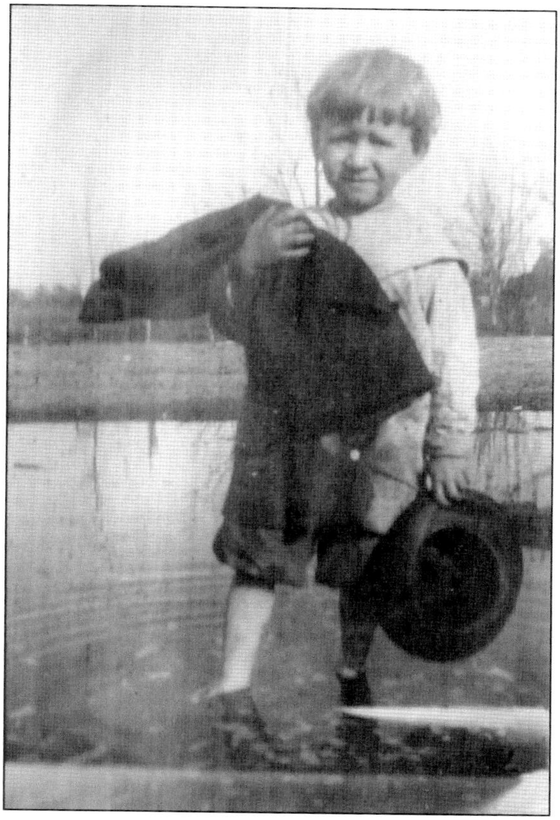

Seen below is another "runaway," Mary Virginia King, or "Ginny," as she was known to all, leaving White House, the family homestead, with her suitcase. She is walking on an oyster bed along the Toogoodoo River at age eight. (Courtesy of Rosanne King Jones.)

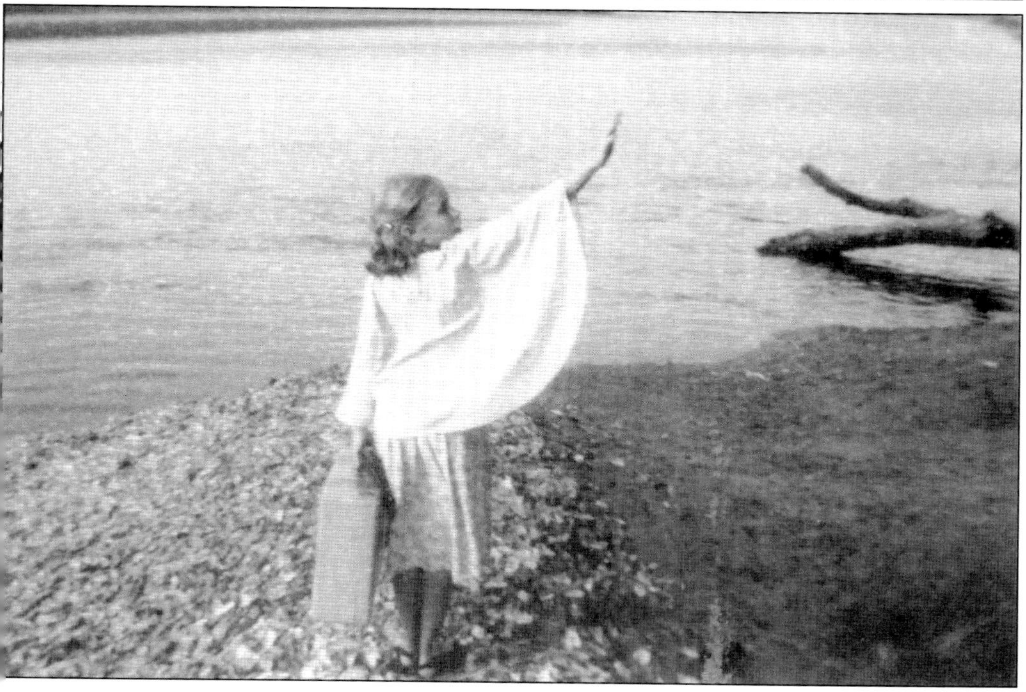

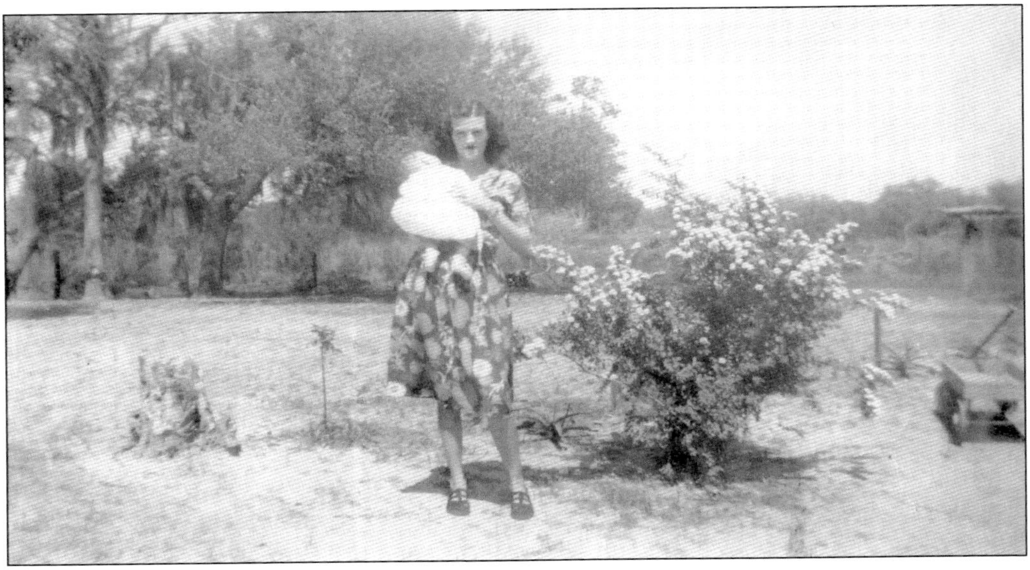

Seen above in the spring of 1947 is Jennie "May" Tumblestone Prentiss holding a young Caroline Prentiss (Jarvis) on their Cherry Tree Farm off Toogoodoo Road. May had been an orphan raised by family in Ridgeville, South Carolina. She then married Edward Burton Prentiss, and they had three children. Different members of the family have resided in St. Paul's Parish, and the farmland is still owned by Prentiss descendants. Below is another picture of Caroline Prentiss (Jarvis) playing in the yard outside the farmhouse in the late 1940s. (Both, courtesy of Caroline Prentiss Jarvis.)

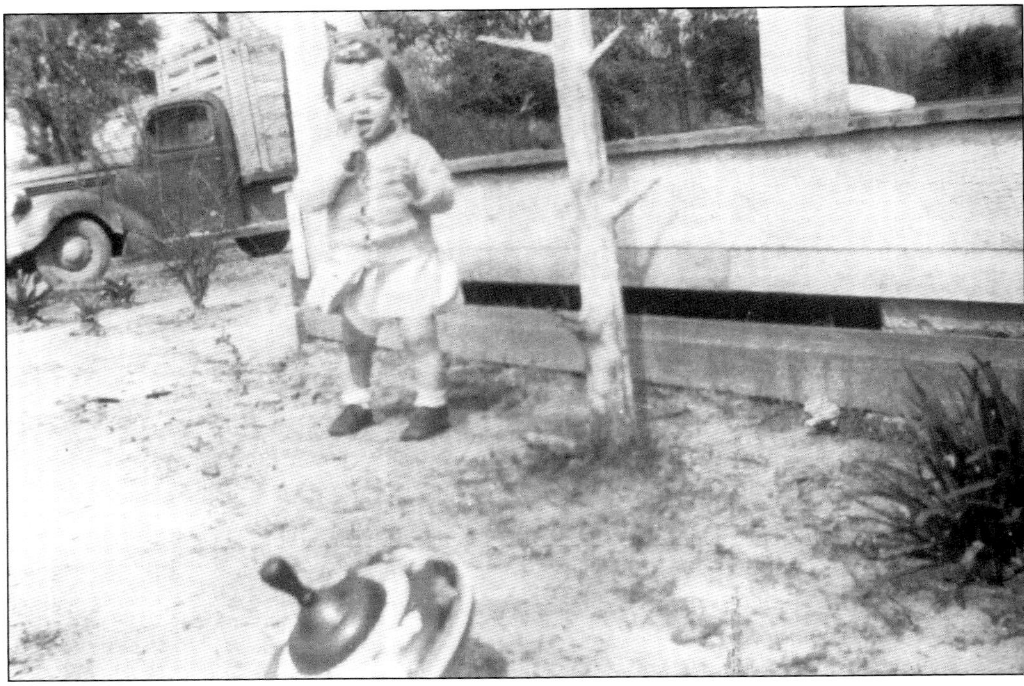

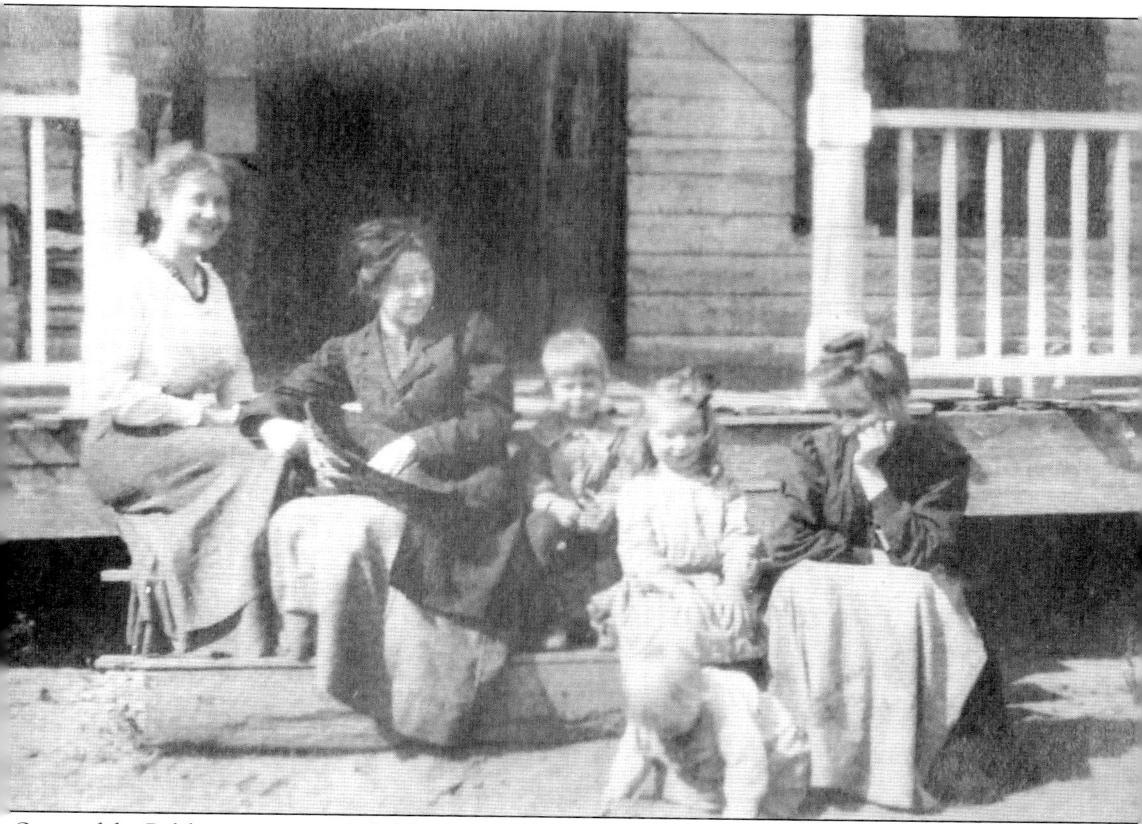

Some of the Behling women and children are sitting on the front porch at their homestead off what is Ethel Post Office Road today. This picture was taken on February 13, 1909. Believed to be in the image are Liz Behling, Hilmer "Hillie" Behling, and Christine Baldwin, along with three children, two of whom are thought to be Hardee and Sadie Butler. (Courtesy of Sarah Roberts.)

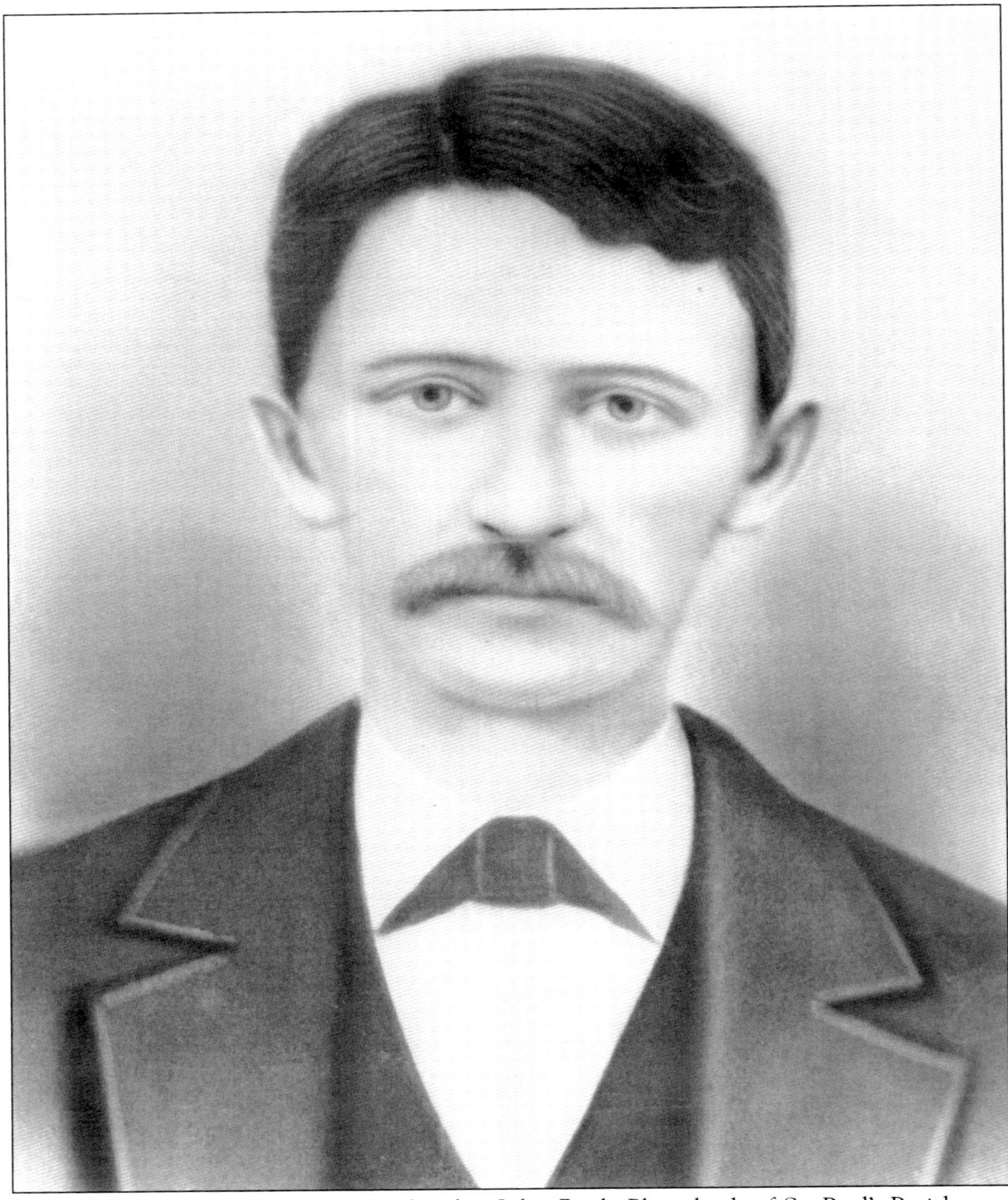

Harmon B. Platt (pictured) and his brother John Boyle Platt, both of St. Paul's Parish, saw Confederate service with the St. Paul's Rangers (Scott's Company), a unit that was part of the 1st (Martin's) South Carolina Mounted Militia Regiment. Both men survived the war and raised their families in the parish. (Courtesy of Elizabeth "Betsy" Waters.)

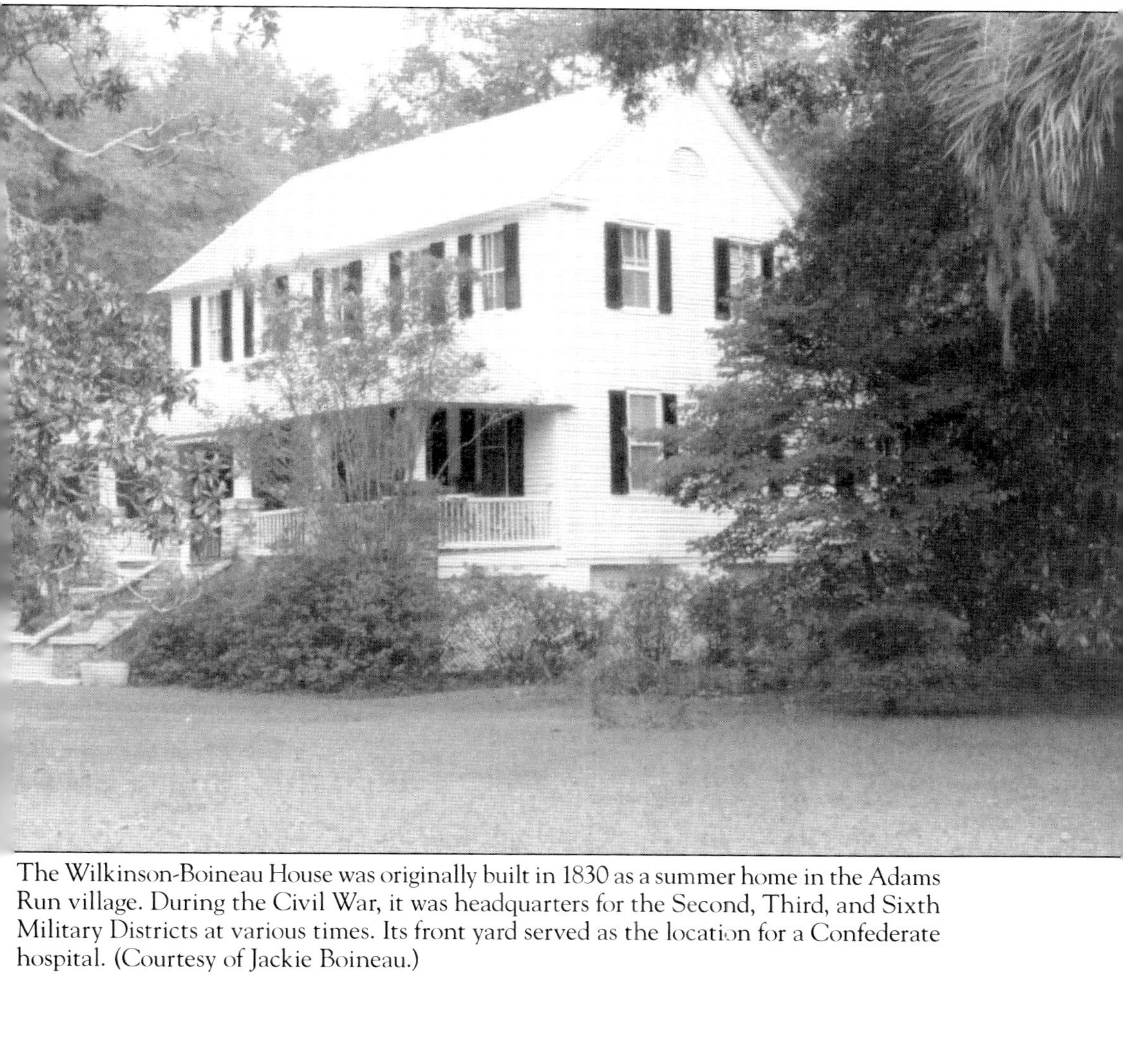

The Wilkinson-Boineau House was originally built in 1830 as a summer home in the Adams Run village. During the Civil War, it was headquarters for the Second, Third, and Sixth Military Districts at various times. Its front yard served as the location for a Confederate hospital. (Courtesy of Jackie Boineau.)

MEGGETT'S NEWS NOTES.

To The Press and Standard: It seems as tho' our correspondant from Meggett has completely gone out of existance, and now too is a very good time of the year, as everything seems to be getting back to the old festive way: but am sorry to say that the farmers all look pretty blue, over their dear little cabbage, which seems to be badly hurt by the cold wave.

Quite a number of the folks around enjoyed a "surprise party" at the home of Mr and Mrs Carlton, on the evening of Feb. 2nd. Several interesting games were played such as pinning on the "donkeys," "snake in the grass" etc. Quite a dainty supper was served at about half after ten o'clock, but the crowd was not willing to return to their homes until the small hours, thus proving that it was thoroughly enjoyed.

The latest "sport" for the young ladies of this section seems to be horse-back riding and boatrowing; Well I dare say its good exercise any way. And they all seem to enjoy it too.

Miss Claude Wilkinson entertained a number of friends at her beautiful home on Toogoodoo on Friday evening at an oyster roast a very enjoyabale time was spent and the memory of it will certainly be a pleasant recollection to all who attended.

Quite an original idea was displayed in hiding the prizes in pretty little clumps of bushes out doors, and letting the guests search for them. The names of the lucky ones may be mentioned as follows; Misses Claude Wilkinson and Liz Behling, and Mess Allston and Ellis of Adams Run.

We were pleased to see Dick Prentiss of Charleston at the party.

Misses Carr and Baldwin of Orangeburg, are the guests of Mrs Carr of Meggett.

Miss Hillie Behling spent part of last week pleasantly with Mrs Butler of Ethel.

Miss Clara Hethington spent a few days in the "city by the sea."

Miss Leila LaRoche has been visiting "on the point."

All of the young folks are still looking forward to the dance that the Masons have promised to give in the new Masonic hall at Meggett.

With best wishes to The Press and Standard.

"Sunshine."

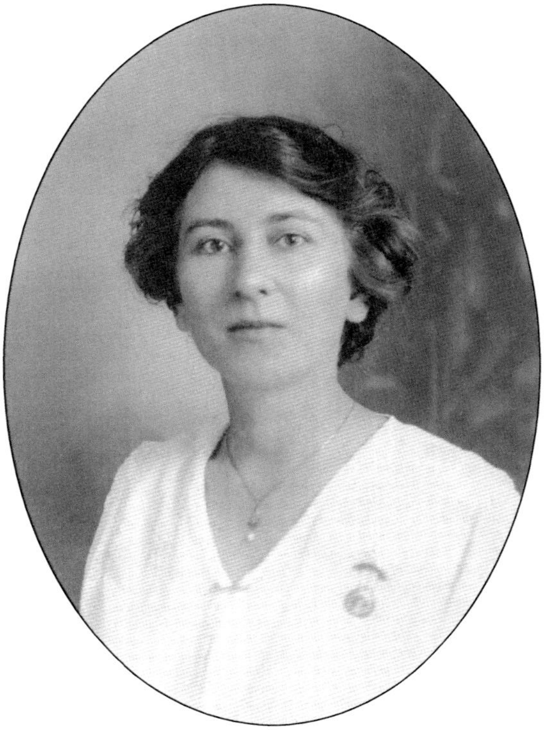

Elizabeth "Liz" Behling was the daughter of Luder Cadow Behling, and Abigail Sarah LaRoche, or "Ms. Abbie." Behling was born on April 7, 1890, on Rescue Plantation in Meggett, and died on June 18, 1974. She married twice during her lifetime: to Chester Lee Smoak in 1912 and to Elisha Calvin Bunch in 1935. Behling was perhaps known best as being a mail carrier for many years for the Meggett Post Office. In addition, she was an author who wrote a byline in the local newspaper under the pen name "Ms. Sunshine." In her column, Behling wrote about the goings-on in town, the weather, and other important events. At left is a copy of one of her articles. (Above, courtesy of Sarah Roberts; left, courtesy of Judy and Reaves Bair.)

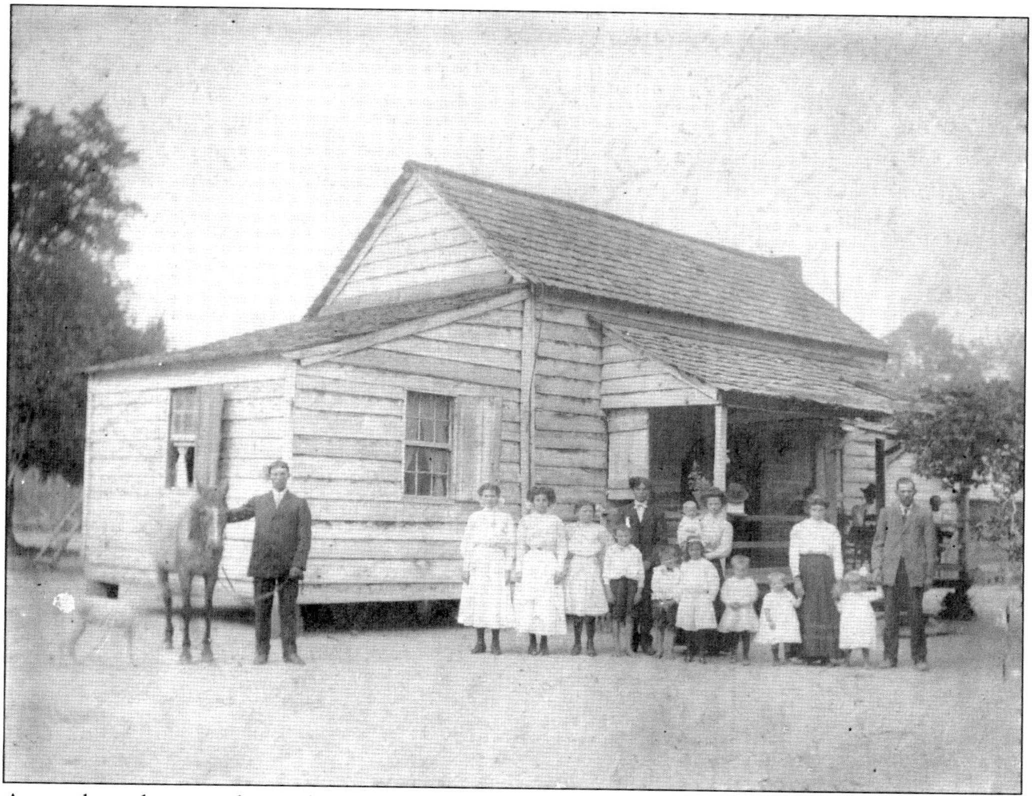

A traveling photographer is thought to have taken this picture of William Stephen and Jane Edmondson Weeks's family, of the Osborn/Parker's Ferry area, outside of their home around 1910. William Weeks (1859–1940) was a local farmer and lumberman as well as a local magistrate. Pictured are, from left to right, (first row) Henry, Sarah, Ruth, Enie, Harrey, Quilly, unidentified girl, Lottie, Ernestine, Janie Savilla Edmondson, Lucille, and William Steven; (second row) Wilkie, Margaret (child), and Elizabeth. Henry R. Weeks (far left, holding the horse reins), son of William Stephen and Jane, was born in 1898 and served in the US Navy during World War I and afterward. Henry is seen at right in his naval portrait. (Both, courtesy of Elizabeth "Betsy" Waters.)

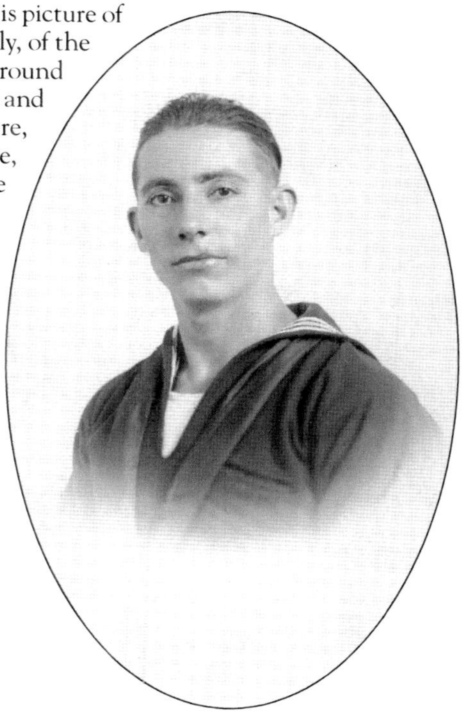

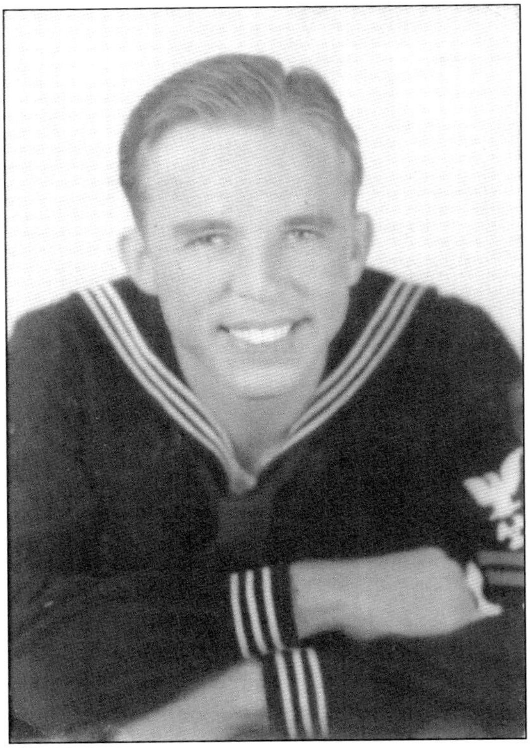

Milton Monroe Muckenfuss is pictured in his formal naval portrait around 1943. Muckenfuss served in the Navy for five years and started as a third-class machinist mate serving on the USS *Jenks* (DE-665), which participated in the first capture of a German submarine, *U-505*, on June 4, 1944. Milton was raised by Claudia "Claude" and Dr. R.R. Prentiss after his mother tragically died when he was seven. Below is a V-mail that was sent to his parents on May 1, 1945. All mail was censored by the government, meaning most messages home were very short and lacked detail; this was very frustrating to loved ones at home wanting news of their family members. (Both, courtesy of Marsha Blandenburg.)

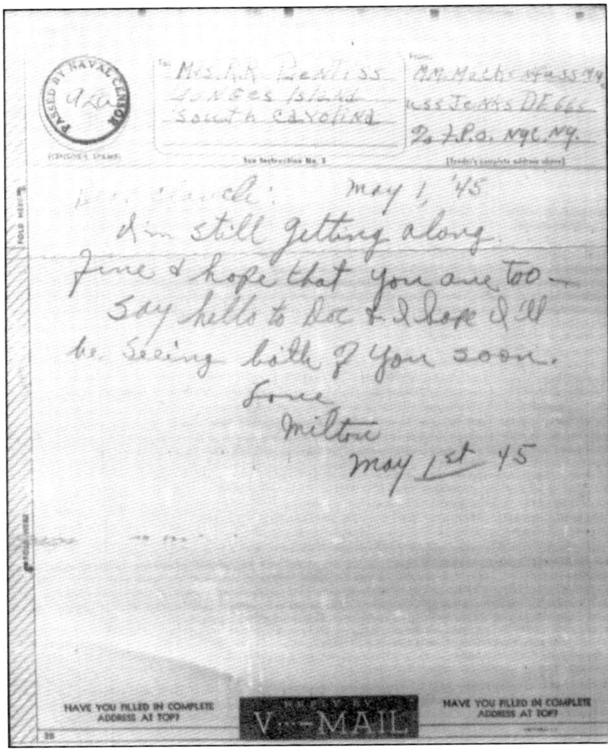

PASSED BY NAVAL CENSOR

May 1, '45

I'm still getting along fine & hope that you are too — Say hello to Doc & I hope I'll be seeing both of you soon.

Love
Milton
May 1st 45

HAVE YOU FILLED IN COMPLETE ADDRESS AT TOP?

V—MAIL

HAVE YOU FILLED IN COMPLETE ADDRESS AT TOP?

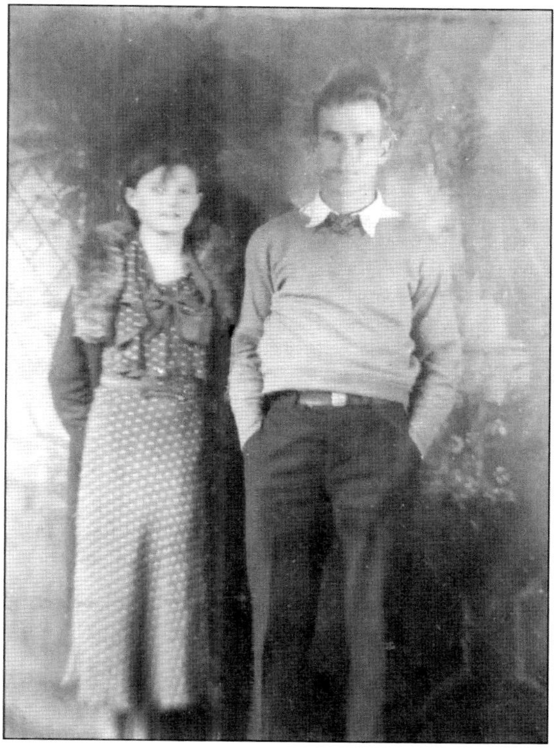

Pictured on their wedding day, July 18, 1935, are Edward Burton Prentiss and his bride, Jennie "May" Tumblestone. The couple, who met in Ridgeville, South Carolina, while Prentiss had been on a hunting trip, married in Charleston and went on to have three children and a long life together in the parish. As a young man 11 years earlier, Edward Prentiss had written this note, or "musing" (seen below), while standing under a tree on his Cherry Tree Farm off Toogoodoo Road wondering what his life would be like 10 years from then. (Both, courtesy of Caroline Prentiss Jarvis.)

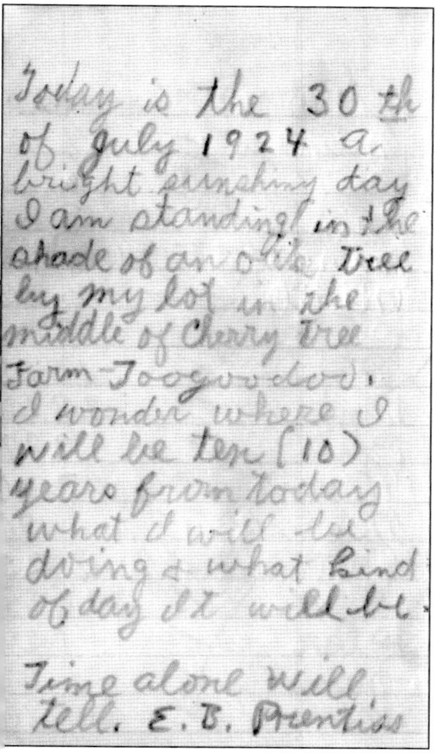

Today is the 30th of July 1924 a bright sunshiny day I am standing in the shade of an oak Tree by my lot in the middle of Cherry Tree Farm-Toogoodoo. I wonder where I will be ten (10) years from today what I will be doing & what kind of day It will be.

Time alone will tell. E. B. Prentiss

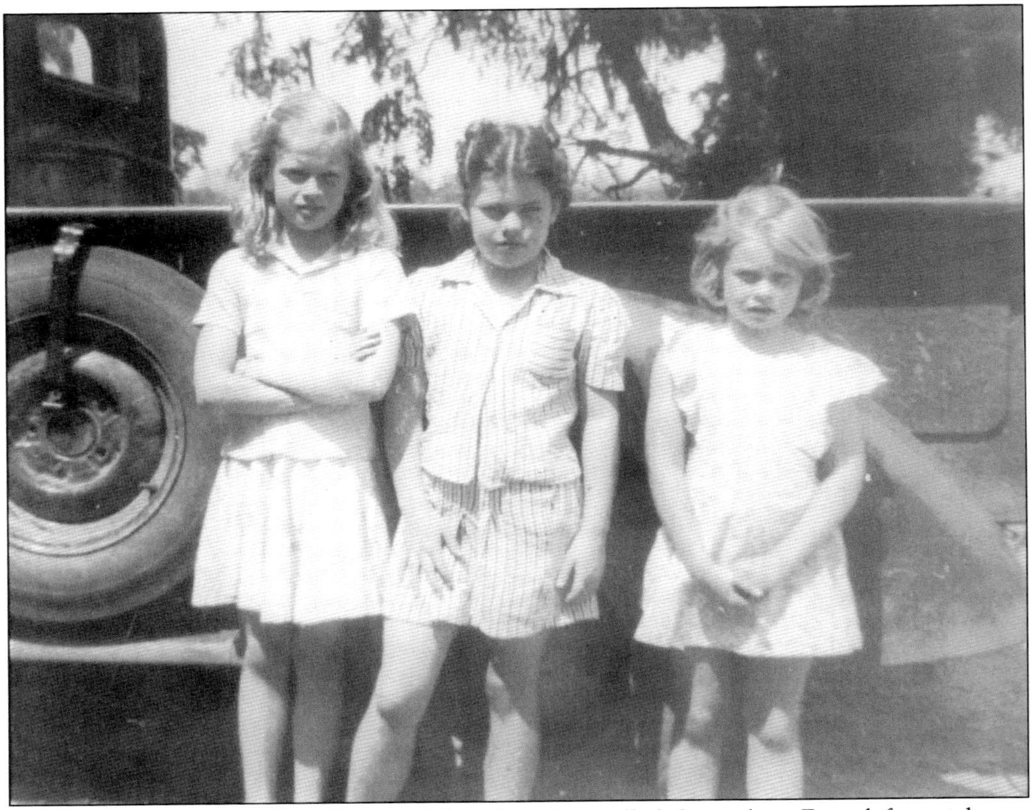

Seen above in 1945 are the "Queens of King," as they called themselves. From left to right are Betty, Harriett, and Mary Virginia "Ginny" posing in front of the family pickup truck outside of White House, the King homestead on the Toogoodoo River. Below is a picture of White House, where generations of Kings grew up. The girls are posing outside again along with an unidentified cousin. (Both, courtesy of Rosanne King Jones.)

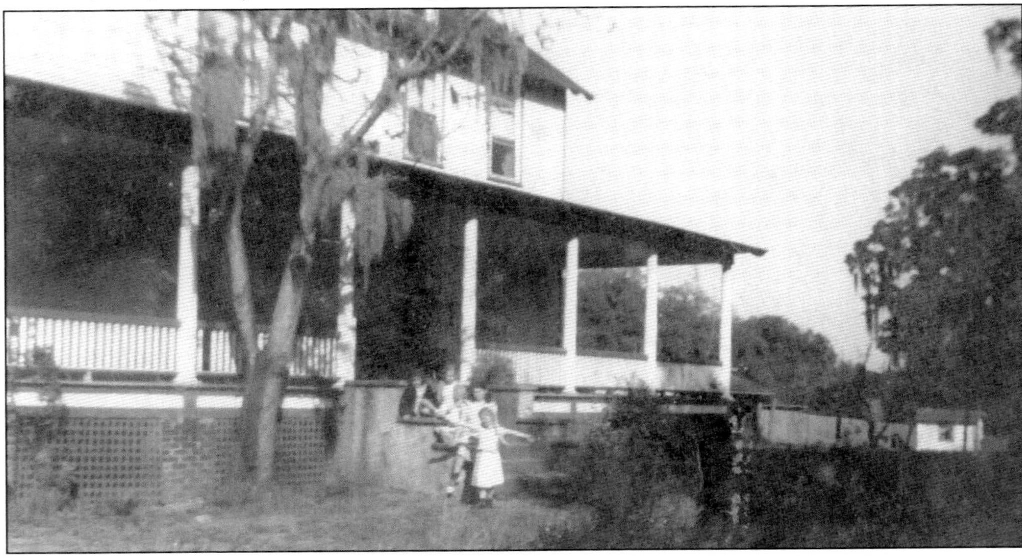

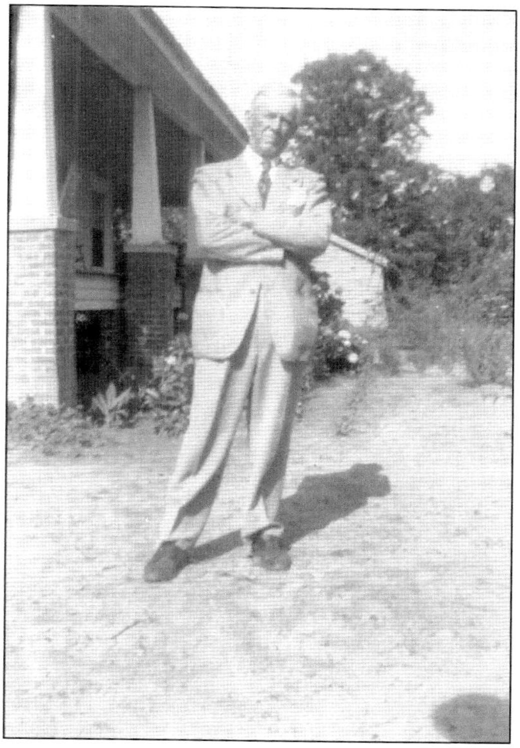

John Jenkins Prentiss, born on May 28, 1879 in Jacksonboro, South Carolina, poses outside of his home in St. Paul's Parish in a "typical stance." Prentiss married his sweetheart, "Macie," on April 18, 1900. Seen below is their wedding invitation, which is a prized piece of memorabilia for family members. The event took place on a Wednesday evening at 8:30 p.m. at "Toogoodoo." (Both, courtesy of Caroline Jarvis.)

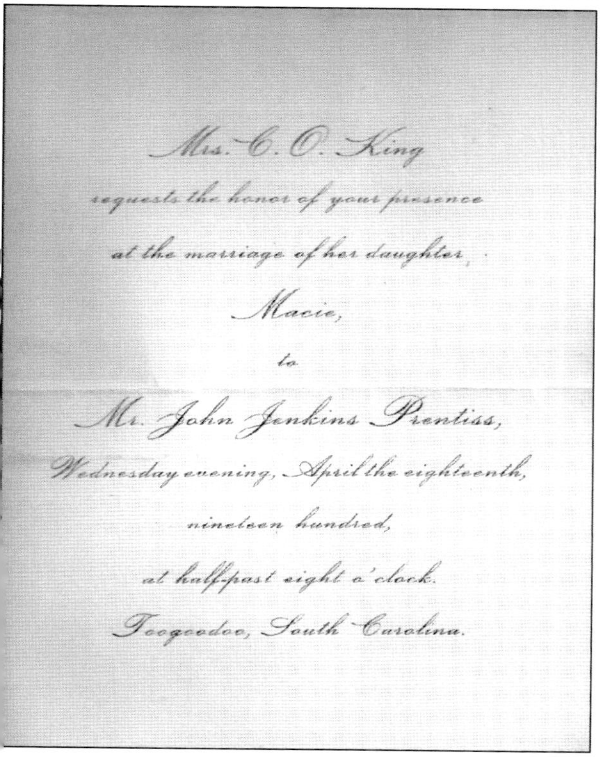

Mrs. C. O. King
requests the honor of your presence
at the marriage of her daughter,
Macie,
to
Mr. John Jenkins Prentiss,
Wednesday evening, April the eighteenth,
nineteen hundred,
at half-past eight o'clock.
Toogoodoo, South Carolina.

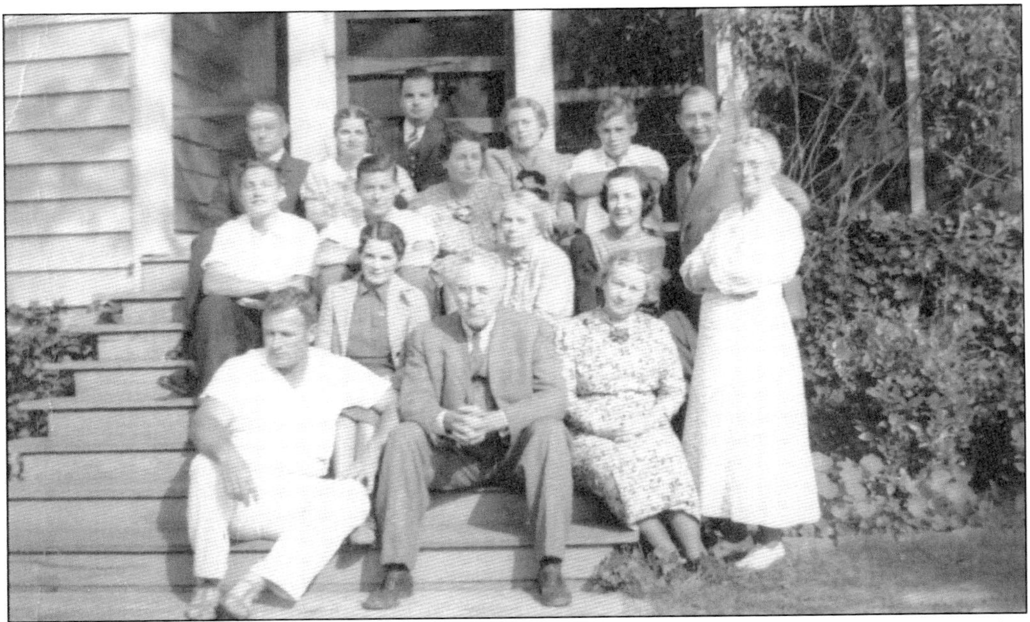

Members of the Geraty family and friends sit on the front porch of their homestead one Sunday afternoon around 1933 or 1934. From left to right are (first row) William Geraty, John William Geraty, Marie Geraty, and Helen Cooper Geraty; (second row) Weston "Duck" Harrison Geraty, Mel Poyle, and unidentified; (third row) all are unidentified; (fourth row) unidentified, Anna Geraty, Willard Silcox, Helen Ann Geraty, Sallie Geraty, and Steele Geraty. (Courtesy of William "Billy" Geraty.)

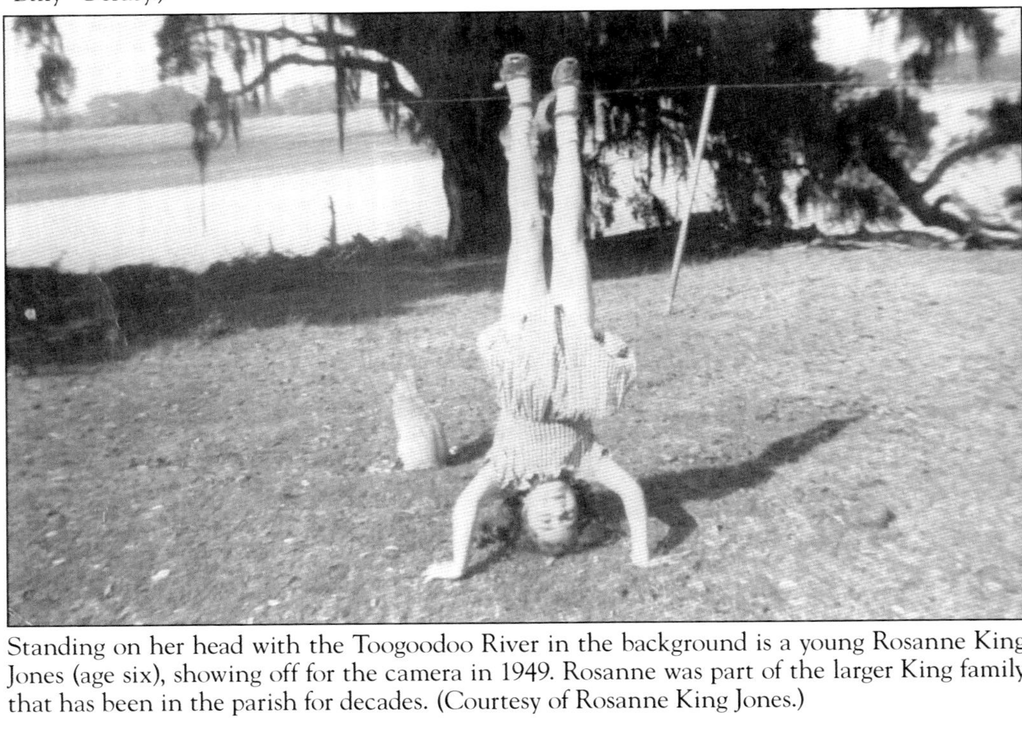

Standing on her head with the Toogoodoo River in the background is a young Rosanne King Jones (age six), showing off for the camera in 1949. Rosanne was part of the larger King family that has been in the parish for decades. (Courtesy of Rosanne King Jones.)

Pictured at right is "Ms. Emma" holding a young William "Billy" Christopher Geraty in 1927 outside of the Geraty home on Yonges Island. Emma was a young African American woman who helped raise many of the Geraty children. She is trying to get young Billy to look at the photographer, but he is clearly not interested at this time. Seen below in the late 1920s are Weston "Duck" Harrison Geraty with her young sons William (left) and Charlie, who was four years older than William. (Both, courtesy of William "Billy" Geraty.)

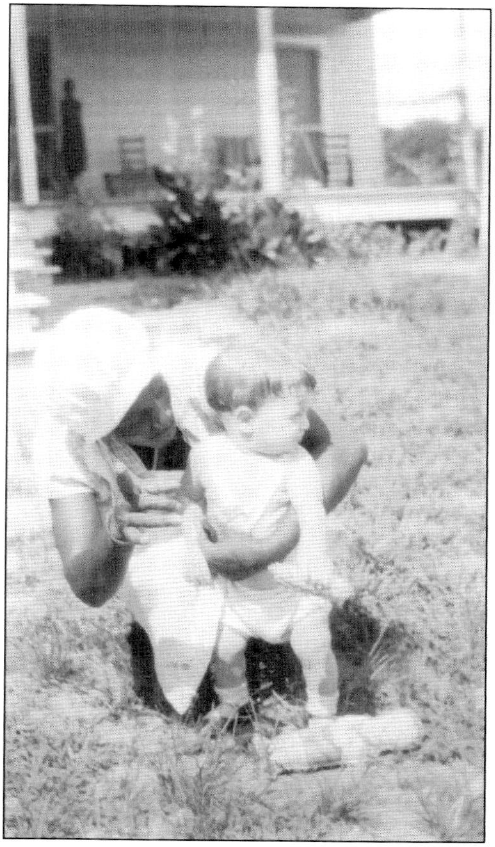

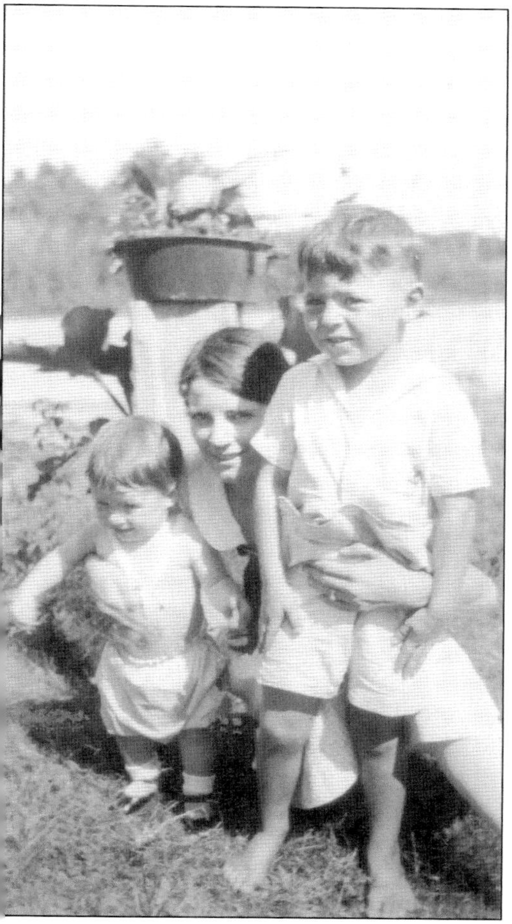

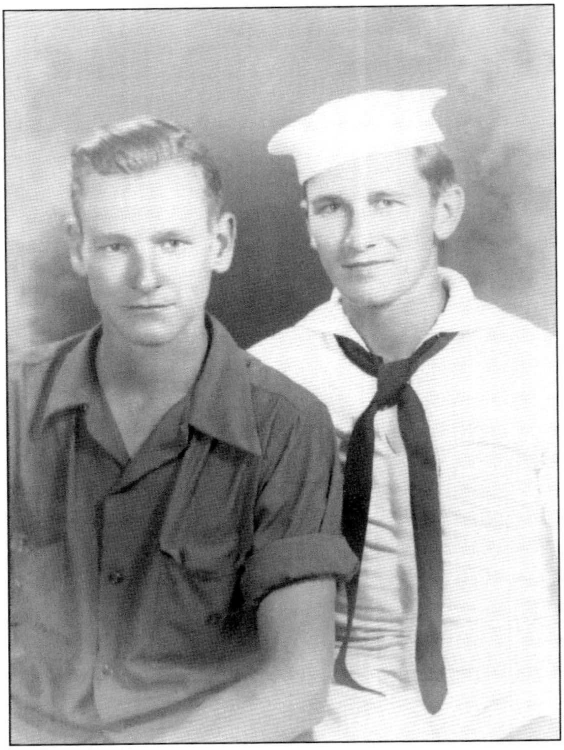

In the photograph at left taken in 1941 are two parish brothers—Joseph "Joe" Prentiss Anderson (left) and Ellison "Ellie" Capers Anderson. Known as "Josie" to most, Joe started his plumbing, electrical, and general contracting business in the mid-1950s. He added well-drilling a few years later, and his company, J.P. Anderson Well Drilling (Anderson Will Drilling), is still active today. Joe is fondly remembered by many for his love of animals, in particular the white squirrels on Yonges Island. The white squirrels on Yonges Island are believed to have been brought to the area by Charles Geraty many years ago. They are not albino but a variant of the common gray squirrel. At his death in 1992, Joe had successfully raised 25 adult white squirrels in his squirrel house, helping to spare them from extinction. (Both, courtesy of Linda Allston.)

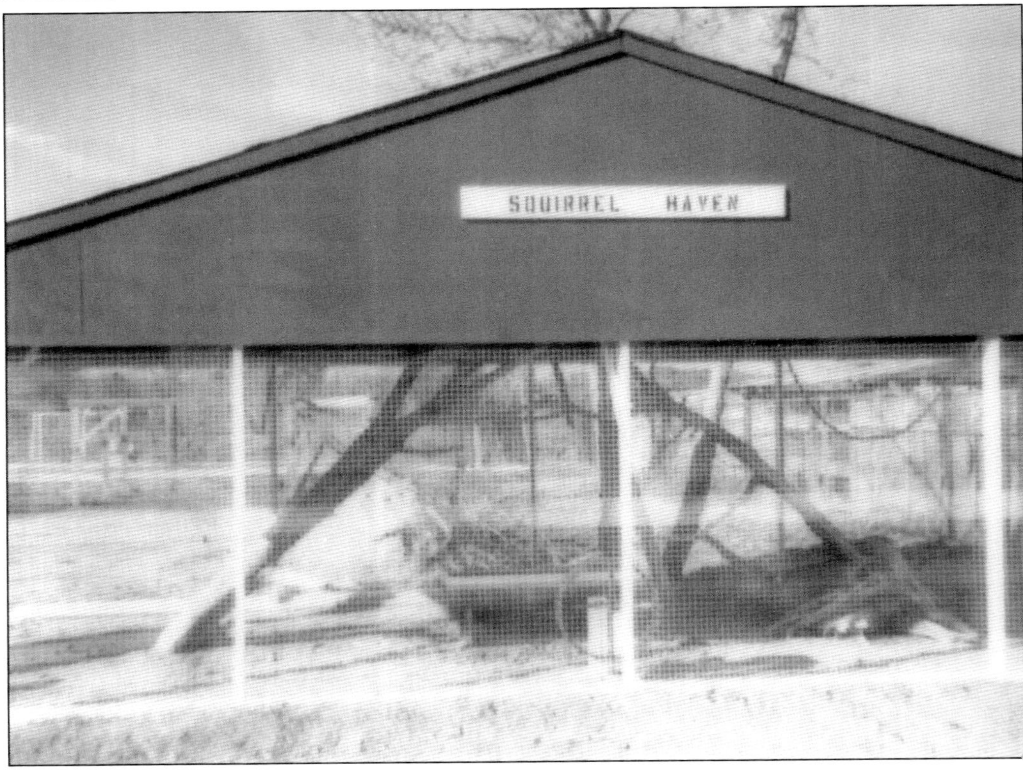

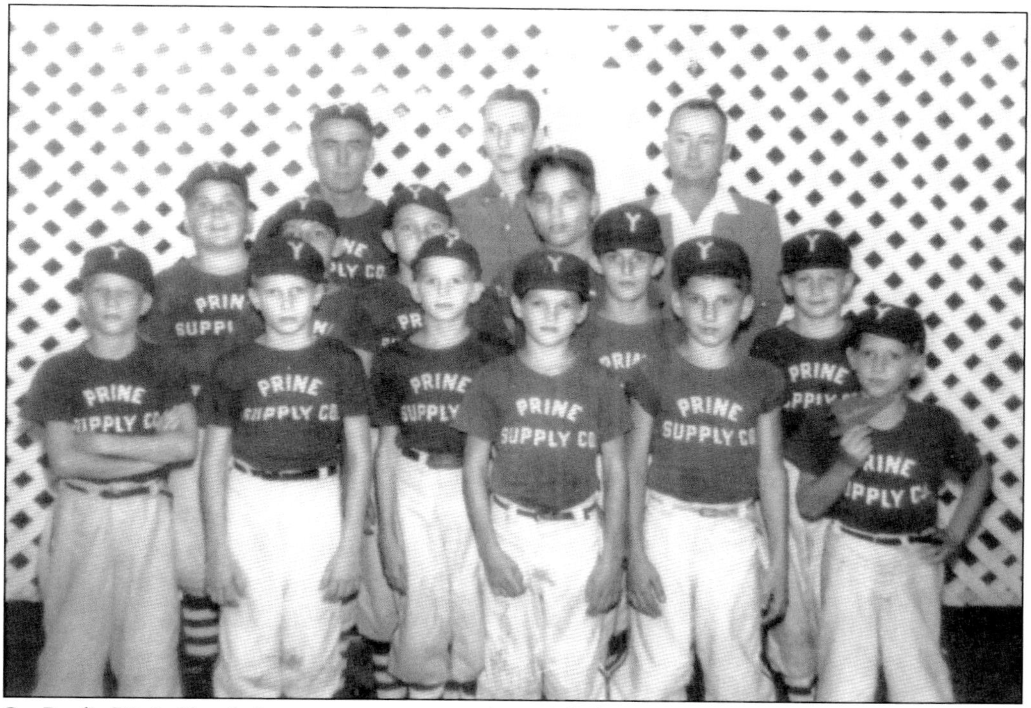

St. Paul's Dixie Youth League was organized in 1957 by Joe Prine, Prine's Supply, and other local business owners. The league consisted of four teams—the Braves, Cardinals, Red Sox, and Yankees. They played at the old St. Paul's High School baseball field on Yonges Island. From left to right are (first row) Timmy Sanders, Pat Sistare, and Teddy Smoak; (second row) Edward Knisley, Robby Perry, Andy Golden, Mike Carroll, and Lane Chaplin; (third row) all unidentified; (fourth row) Joe Prine, assistant coach Roy Smoak, and assistant coach Rodell Smoak. (Courtesy of Bobby Garvin.)

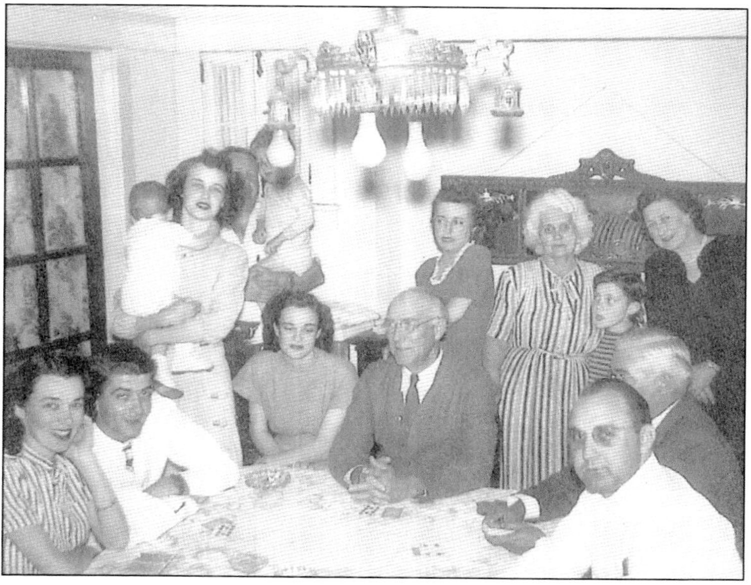

Photographer Bubba Welch captured a group of parish residents and friends at a dinner party in the mid- to late 1940s. At the table are, from left to right, Sissy Welch, Sleepy Chaplin, Ester Ann holding baby Lance, an unidentified man and child, Dot, Bagot's Dad, Esther Jenkins, Bagot's Mom, Bill Searson, Gladys Rodgers, Gladys's second husband (with white hair), and Bagot Searson. (Courtesy of Jackie Boineau.)